Female STEM Pioneers who survived the Holocaust

Female STEM Pioneers who survived the Holocaust

Victoria Van Orden Martínez

PEN & SWORD
HISTORY

First published in Great Britain in 2026 by
Pen & Sword History
An imprint of Pen & Sword Books Limited
Yorkshire – Philadelphia

ISBN 978 1 03612 576 9

A CIP catalogue record for this book is available from the British Library.

Typeset by Mac Style
Printed in the UK by CPI Group (UK) Ltd, Croydon, CR0 4YY.

FSC www.fsc.org
MIX
Paper | Supporting responsible forestry
FSC® C013604

The Publisher's authorised representative in the EU for product safety is Authorised Rep Compliance Ltd., Ground Floor, 71 Lower Baggot Street, Dublin D02 P593, Ireland.
www.arccompliance.com

For a complete list of Pen & Sword titles please contact:

PEN & SWORD BOOKS LIMITED
47 Church Street, Barnsley, South Yorkshire, S70 2AS, England
E-mail: enquiries@pen-and-sword.co.uk
Website: www.pen-and-sword.co.uk
or
PEN AND SWORD BOOKS
1950 Lawrence Road, Havertown, PA 19083, USA
E-mail: uspen-and-sword@casematepublishers.com
Website: www.penandswordbooks.com

Contents

Acknowledgements

Researching and writing this book has been a labour of love that I have had the pleasure to share with so many amazing people, all of whom I owe my sincere thanks. First, I am grateful to my husband David Martínez Ortega and our two children, whose patience as I've hidden myself away at home to write has been unending. Second, to my incredible scientific and critical editor, Jackie Duffin, whose matchless expertise and experience were invaluable in terms of improving the manuscript, both scientifically and stylistically. Of course, I don't know what I would have done without the support of my two sisters-of-the-heart, Amy Cohen and Marci Peters, who not only took the time to read the draft manuscript but even joined me on some research visits! Thanks to my friend Roman Wroblewski, who knew and worked with Halina Neujahr and has given me support and insight into her and so much more over the years. Because this book is the product of research I have conducted in the course of my academic work, I also express my sincere thanks to my colleagues near and far, who have listened to countless seminars and presentations on my research, read much more formally written texts on the topic, and provided invaluable feedback on all of these outputs that ultimately contributed to this book. I am especially grateful to my late mentor and friend Paul Levine and to Malin Thor Tureby, who believed in me and guided me into the world of Swedish academia, and to Henrik Rosengren and Ulf Zander at Lund University for making my dream of working at this wonderful university come true.

I am also deeply indebted to the many librarians – especially at Linköping University and Lund University – who have helped me access many books and other resources during my research. Over the years, I have visited a multitude of archives and libraries to conduct research for this book and have benefited from the digital collections made available by many others. As a non-residential fellow of the University of Southern California (USC) Visual

History Archive, I have had the privilege of unlimited access to the collection of video testimonies given by more than 52,000 Holocaust survivors. This has allowed me to spend time watching and re-watching the testimonies of some of the women featured in this book and many others who are not. I especially want to thank Jennifer Rodgers, formerly of the USC Shoah Foundation, for her help and consideration. Likewise, I am grateful to the Yale Fortunoff Video Archive for Holocaust Testimonies and Stephen Naron for access to the collection. My special thanks to the archivists at the Lund University Library's Special Collections, the Swedish National Archives (*Riksarkivet*), the Nordic Museum (*Nordiska museet*), the Karolinska Institute (*Karolinska Institutet*), and the Royal Institute of Technology (*Kungliga Tekniska högskolan*). Likewise, I am grateful for the hospitality of archivists Kendall Martin at the UT Southwestern Special Collections Library and Archives, Felicia Williamson at the Dallas Holocaust and Human Rights Museum, and the librarians at the J. Erik Jonsson Central Library of the Dallas Public Libraries. I have utilised so many archives with online repositories that it's impossible to name them all. However, I would like to make special mention of the Arolsen Archives – International Center on Nazi Persecution, the United States Holocaust Memorial Museum, Yad Vashem – the World Holocaust Remembrance Center, the Marine Biological Library at the University of Chicago, the Wiener Holocaust Library, the Imperial War Museum, and the *Svenska tidningar* (Swedish newspapers) database. I am grateful to all the individuals and institutions that have given permission to publish images in their collections, including the Bernard Becker Medical Library at Washington University (with special thanks to Philip Skroska), the Historical Archive of the University of Turin, the Dallas Holocaust and Human Rights Museum, *The Chicago Tribune*, the UT Southwestern Special Collections Library and Archives, Öreryds Hembygdsförening, the Karolinska Institute, the USC Shoah Foundation (with special thanks to Rachel Herman), Roman Wroblewski, and Jacalyn Duffin.

Finally, a big thank you to Pen & Sword Books for publishing this book and helping me to share these women and their accomplishments with the world. I'm so grateful to whomever suggested me to Lucy May, my original commissioning editor (as well as to Lucy, of course), to Sarah-Beth Watkins, who took over after Lucy, the editors who helped refine the manuscript, and the designers and others who made it a beautiful book.

Introduction

The origins of this book can be traced back to one woman: Halina Neujahr. My first glimpse of Halina came in 2017, while I was researching Jewish and non-Jewish Polish survivors of Nazi persecution who came to Sweden as 'repatriates' – refugees expected to stay in the country temporarily – in the spring and summer of 1945, right at the very end of the Second World War. I was new to Sweden then, having only recently arrived with my family.

While exploring my new home and its history, I was surprised to discover that around 1,000 Jewish and non-Jewish Polish survivors of concentration and labour camps had recuperated in a refugee camp not far from where we lived, in a quaint village called Öreryd. According to the little that was written about these people at that time, they had all gone away after 1946; to where, nobody was quite sure. Probably, it was speculated, to North America or Palestine/Israel. Unsatisfied with this tidy ending, I decided to research who some of these individuals were and what became of them. It didn't take me long to find Halina Neujahr, not in North America or Israel, but in Sweden.[1]

When I write that I found her, I mean that I found traces of her life and career, since she had died in 2006. To be honest, she was not hard to find. It was a case of lost in plain sight, at least in terms of the history of the Polish survivors who had found refuge in the Öreryd refugee camp. Far from disappearing into thin air, Halina had not only stayed in Sweden but had become in 1960 the first woman in Sweden to earn a Doctor of Technology degree in chemistry. Earning that distinction came with the further distinction of being only the third woman in Sweden to earn *any* Doctor of Technology degree. She was the only one among the three who was not born in Sweden, and the other two earned their degrees not long before she did. Halina had not only survived the Holocaust, which robbed her of most of her family, but she also overcame incredible obstacles as a foreigner, Jewish woman, and refugee to enter a male-dominated field in Sweden and accomplish what few Swedish women had done at that time.

My research on Halina brought me closer to her incredible life story. Sitting in an archive in Stockholm, browsing her personal photo albums, I felt that I was sitting with her, listening to her tell me about the extreme ups and downs of her personal and professional life. At the same time, my research was also informing me that while Halina Neujahr's story was rare, it was not singular. There were more women like her; women who survived the Holocaust and other forms of Nazi racial and ethnic persecution and went on to become pioneers in male-dominated STEM (Science, Technology, Engineering, Math) professions, many outside of their native countries.

That these women survived the Nazis' racial and ethnic policies is nothing short of astonishing. As is well known, the Nazi agenda was total annihilation of all Jews. Even having failed to win the war, the Nazis had nearly achieved their genocidal ambitions in countries like Poland, where almost the entire prewar Jewish population of more than three million had been murdered. This figure accounts for almost half of the estimated six million European Jews murdered by the Nazis. Throughout Europe, it is estimated that two of every three Jews had been killed between 1933 and 1945.[2] Others subjected to Nazi genocide included individuals with mental and physical disabilities, the Roma and Sinti (or 'Gypsies'), and homosexuals.

The Nazis also targeted Slavic people – such as Poles, Czechs, Slovaks, and Russians – as an inferior race and therefore enemies who deserved enslavement or death. The result was that, in addition to the millions of Jews from Slavic countries who were murdered, the Nazis also imprisoned, executed, and murdered millions more non-Jewish civilians from these countries. The Nazis were particularly concerned with eliminating well-educated individuals, both Jewish and non-Jewish – including professors, politicians, judges, teachers, lawyers, the clergy, and, of course, physicians – due to the influence they had over the masses, whom the Nazis wished to rule without resistance. Of course, where selections at concentration, labour, and death camps were concerned – the moment when an SS official would decide who went to the gas chamber and who would be sent to death-by-labour – women frequently fared worse than men. Visibly pregnant? Death.

Holding a baby or the hand of a child? Death. Too young, old, or physically weak for labour? Death. Jewish woman who just might procreate another Jew if she lives? Death. At least when you were sent to death-by-labour there was a slim chance you would survive. This was a chance fewer women than men would have had because of their gender.

Where survival was concerned, usefulness came in handy. Although the Nazis were quick to dispose of intellectuals, medical professionals sometimes proved useful to them in the conditions of overcrowding, filth, and epidemics that defined the concentration and forced labour camps. Medical forced labour was commonplace in these conditions. Not surprisingly, more men than women were useful to the Nazis in this respect. Advanced medical training was still largely reserved for men in most of prewar Europe. As a result, Jewish and non-Jewish physicians spared from death – at least temporarily – were typically men. But women were far from scarce among forced medical labourers during the Holocaust. Some women worked as doctors based on their prewar qualifications while many others worked as nurses, either based on previous training or by some stroke of luck at selection. I write 'luck' not because they necessarily wanted to do this work but because, as some later reflected, being selected for this work gave them a chance to survive.

Those who were fortunate enough to survive these circumstances were faced with a whole new set of challenges. If you were among the millions of people liberated from Nazi concentration and labour camps in 1944 and 1945, you were likely also counted among the displaced persons (DPs) of the period who were either placed in DP camps operated by the Allied forces, or you were among those set adrift, often searching for lost loved ones. Finding your way personally was difficult, not to mention finding your way professionally. In terms of opportunities for women, conditions had not necessarily improved since the war started. In general, circumstances were bad for most everyone in the countries devastated by the war. Many DPs refused to return to their home countries, where often nothing remained of their families and former lives. For some, creating a new life elsewhere was preferable to trying to recreate what had been lost in their native countries. In other cases, a return 'home' was simply a necessary or even imperative stop before embarking on the often-onerous process of obtaining permission to migrate elsewhere.

For those who had worked in professions in their home countries – not least medical and other scientific fields – before or during the war, emigrating elsewhere after the war often involved discovering that their education, experience, and even their credentials were not recognised. More than a few promising professional careers ended when survivors chose to make new lives outside their home countries after the war. Among the women I have researched, most of those who had successful STEM careers outside their native countries were relatively young and/or had earned their degrees soon before the Nazi occupation of their homelands or after the war in their adopted countries. Most older women or those who had worked as STEM professionals before the war either continued their careers in their home countries or gave them up after emigration.

The women featured in this book were all young, or relatively young, women who encountered their new homelands with a certain vigour, a determination to pursue their futures while also, contrary to expectations, *allowing* their pasts to define them. Rather than rejecting their experiences under Nazi persecution, they embraced them and effectively made them a part of their motivation and drive to overcome the obstacles in front of them. Not only the obstacles inherent in the treatment of incurable diseases and making advances in medical science, but also the obstacles facing women in male-dominated STEM professions.

Many of these obstacles are still in place today, eighty years after the end of the Second World War and the Holocaust. If women in our own time face greater inequality and less recognition in medicine and science than men, we can imagine what it was like during these women's lives. One of the six women featured in this book won the Nobel Prize: neurobiologist Dr Rita Levi-Montalcini, in 1986, in the category of Physiology or Medicine, and that award was shared with a male scientist. This is not to suggest that the other women in this book were deprived of the honour. Rather, it emphasises the fact that, as of 2022, 96.4 per cent of all Nobel laureates in science were men even though an increasing number of women were obtaining degrees in STEM. A primary reason for this discrepancy is that women's scientific work and accomplishments are likely to go unrecognised and uncredited.[3] This is just as true now as it was in the past.

Of the six women featured in this book, only Rita Levi-Montalcini and Polish Canadian chemist Halina Czajkowska have been the dedicated subjects of any published books, at least as far as I have been able to determine. In addition to the books written by Rita herself are some short biographies intended for young readers.[4] Halina Czajkowska's memoirs were published shortly before her death in 2021, thanks to the efforts of Canadian haematologist and medical historian Dr Jacalyn Duffin, who earlier published a series of scientific articles that brought Halina's important contributions to science to light for the first time.[5]

Halina Neujahr was known during her life as both a groundbreaking woman in science and a Holocaust survivor. Yet, what has been written about her consists primarily of decades-old Swedish magazine and newspaper articles and a few contemporary ones. One chapter is dedicated to German-American haematologist Dr Ellen Loeb in the book *Sisters in Sorrow: Voices of Care in the Holocaust* by Roger A. Ritvo and Diane M. Plotkin.[6] However, it consists only of translations of correspondence between Ellen and her family and provides little insight into her life and professional career. That said, it is thanks to Plotkin that oral histories and other materials related to Ellen Loeb are available to researchers through the Dallas Holocaust and Human Rights Museum and the United States Holocaust Memorial Museum. Of the other women, although each made important contributions to science and medicine in the second half of the twentieth century, their legacies are mainly recorded through their scientific publications, in a smattering of newspaper and magazine articles, or as secondary figures in books about their male colleagues. Hungarian Swedish cancer immunologist Dr Eva Klein, for instance, is as internationally renowned as her husband. However, far more has been written about him than has been about her.

I have written this book to contribute to filling the gaps, at least concerning these six women – Dr Rita Levi-Montalcini, Dr Eva Klein, Dr Ellen Loeb, Dr Ruth Pick, Halina Czajkowska, and of course, Dr Halina Neujahr. Many more women like them are still awaiting similar recognition. In different ways, they became STEM pioneers because they shattered glass ceilings, paved the way for other women to enter STEM, and contributed to making discoveries and improvements in their respective fields that changed and/or improved the human condition. But they were pioneers in other respects, too. They broke through other barriers besides gender, including racism, xenophobia, and antisemitism. They turned traumatic experiences of persecution, genocide, and the destruction of their homes and families into life-affirming professional and personal ethics and practice.

While each of these women deserves a full biography, my intention is not to provide a series of exhaustive biographies but rather to weave through their lives and experiences in a way that demonstrates how closely their scientific careers were entangled with their experiences of survival during and immediately following the Second World War and the Holocaust. I have chosen these six women in part because of the diversity of backgrounds and experiences in their home countries of Germany, Italy, Czechoslovakia, Hungary, and Poland, and in Nazi concentration and labour camps, hiding, and exile, as well as in the aftermath. All but one – Halina Czajkowska – were Jewish. Though not subject to genocide, Halina Czajkowska was nonetheless persecuted by the Nazis because, as a Pole, she was considered to be racially inferior. Her story reminds us that other victim groups suffered under Nazi rule.

Despite their relative heterogeneity, each woman gained professional training and experience in part during the Second World War and the Holocaust and as refugees during the early postwar period. Moreover, as refugees, migrants, and eventually citizens in the United States, Canada, and Sweden, they all had to shift from surviving the Nazis to navigating education and careers in male-dominated STEM fields; namely, neurobiology, biochemistry, haematology, cardiology, cancer research, and cancer immunology. While glass ceilings, gender bias, and the other challenges they faced in their scientific careers might have seemed inconsequential compared to Nazi persecution, their stories demonstrate that their accomplishments were not achieved without difficulty and sacrifice.

It has been an honour and a privilege to trace the incredible lives and careers of these six women and to be able to share them in this book. I thank Halina Neujahr for starting me on this journey.

Prologue

'Mrs Klein' and Other STEM Pioneers

In August 1997, Eva and Georg Klein were interviewed by the USC Shoah Foundation as part of an initiative to make video recordings of the experiences of Holocaust survivors. Eva and Georg had survived the Holocaust in Hungary, where they met after the war while attending medical school. They married and settled in Sweden, where each earned a degree in medicine and a PhD and became pioneers in the research field of cancer immunology. At the time of the interviews, both held the title of Professor and were internationally well known and respected medical researchers. Their interviews were conducted individually but on the same day, in the same location, and by the same interviewer. Both were asked questions about their scientific education and accomplishments. Throughout Georg's interview, the interviewer, Jan Walter Levy, consistently referred to Georg as 'Professor Klein'. During Eva's interview, he referred to her as 'Mrs Klein'.

Why is it (still) so easy to forget that women are more than the daughters, wives, and mothers of the men in their lives? Why is it that other categories of identity – Holocaust survivor and scientist, for example – seem to make this temporary amnesia more acute? Is Woman + Holocaust survivor + Scientist harder to compute than Man + Anything? In the USC Shoah Foundation interviews, whereas Georg was easily 'man' and 'Holocaust survivor' and 'professor', Eva was more summarily 'wife' and 'Holocaust survivor' with a touch of 'how did you manage also being a scientist?' on the side.[1]

This book rests on the seemingly obvious premise that women – like men – are no more or less one aspect of their identity than another. We are all composed of multiple aspects of identity and experience that exist within us in both harmonious and disharmonious ways. The six women featured in this book were shaped by their experiences during the Second World War and the Holocaust, which in turn shaped their subsequent lives and careers in STEM. During their respective careers in haematology, cancer immunology, cancer research, neurobiology, biochemistry, and cardiology, Dr Ellen Loeb, Dr Eva Klein, Halina Czajkowska, Dr Rita Levi-Montalcini, Dr Halina Neujahr, and Dr Ruth Pick saved lives and improved healthcare and sanitation through their scientific work and discoveries. Their accomplishments were achieved both despite and because of the Holocaust and the Second World War.

As many Holocaust survivors have said, surviving the Holocaust was largely a matter of luck. At the same time, many survivors who worked in medical forced labour – including several of the women in this book – also credit their survival to making themselves nearly indispensable through this work. While the medical careers of many women who survived the Nazis ended with the war, the women in this book continued pursuing or began careers in medicine and science. After liberation, they carried with them all their metaphorical baggage – emotional and physical trauma, loss, and destruction. But instead of allowing it to destroy them, they harnessed their experiences and transformed them into patience, resilience, determination, knowledge, and dedication to contributing to humanity. What follows is a brief introduction to each of the six women through snapshots of their contributions to medicine and science.

Dr Ellen Loeb: The Strength to Treat Something So Fragile

English writer Sir Osbert Sitwell eloquently described blood as 'that fragile scarlet tree we carry within us'.[2] For Sitwell, blood was a source not only of life but also of wisdom. It was the same for Dr Ellen Loeb, whose medical career as a haematologist was dedicated to the study and treatment of blood diseases and disorders, such as leukaemia, lymphoma, and haemophilia. The fragility of Sitwell's scarlet tree was apparent to her in every patient suffering from incurable diseases and disorders of the blood.

In 1955, early in Ellen's career, the Leukemia and Lymphoma Society reported that 'leukemia is 100% fatal' despite the advent and increased use of chemotherapy.[3] In 1960, life expectancy for people with severe haemophilia was less than 20 years – an improvement over the previous decade, when it was 13 years or less.[4] In response to these grim statistics, Ellen harnessed the power of blood as a source of life and wisdom. Working against certain death, she focused her efforts on extending the length and quality of life of her patients. Following Ellen's death, one journalist wrote that she had 'treated thousands through her distinguished career and gave the living a cherished hold on life, a ray of sunshine'.[5] She did this in part by being among the first physicians to test the latest cutting-edge treatments for leukaemia and other blood diseases, which she carefully documented and shared with colleagues around the world through journal articles and conference presentations. 'She was a clinician with a keen interest and talent for the science of medicine and played a major role in the design and implementation of several trials of new treatment strategies for leukemia and cancer,' recalled a physician who worked with her.[6]

As the daughter of a prominent German physician, Ellen had early ambitions to practise medicine. Any chance of that was seemingly shattered when the Second World War began and she and her family were deported by the Nazis. A combination of a determination to survive and brave assertiveness in the face of death led her to work as a forced medical labourer in several Nazi camps. While assisting German doctors in the Theresienstadt Ghetto in Czechoslovakia, her interest in the application of innovative new medical treatments for blood diseases was ignited.[7] During her career after the war, she then contributed to developing new treatments for leukaemia that are still used today. She also made prescient decisions about *not* using certain treatments, saving the lives of many patients as a result.

As important as her contributions to advancing the treatment of and searching for cures for blood diseases and disorders was the way she cared *about* and not just *for* her patients. 'She was a humanitarian who did more than treat her patients,' recalled her medical partner's son, also a physician who worked with her. 'She was attentive to their individual happiness and well-being while in her care.'[8] Ellen's personal ethic of care was especially important with the many very young children she treated. This, too, could be traced back to her experience as a forced medical labourer during the Holocaust. Recalling her experiences in the Westerbork transit camp in the Netherlands – from where Jewish prisoners were sent to almost certain death in the gas chambers at Auschwitz – Ellen wrote at length about the children placed in her care.

> Little ones, many under three years of age, didn't even know their names. They had no parents and many were too weak to talk. But we tried to do the best we could. You just cannot believe what went on. I could write for hours about their misery.[9]

Years later, working at the Wadley Research Institute and Blood Center in Dallas, Texas, Ellen continued to do the best she could for the children she treated, going above and beyond the call of a physician. Children from all over Texas were brought to Wadley, where they often stayed for long periods. 'Ellen worked with volunteers who acted as part-time surrogate parents,' recalled a colleague. 'She also arranged for them to have something of interest or beauty in their room.'[10]

Part of Ellen's great strength was her ability to handle something as fragile as blood, from which she expanded wisdom about how to treat blood diseases and disorders and extended the length and quality of her patients' lives. Today, the life expectancy of individuals with severe haemophilia has increased dramatically when patients receive proper treatment, to only around ten years less than individuals without haemophilia. As for cancers of the blood – or haematologic malignancies – such as leukaemia and lymphoma, they are currently the fifth most common type of cancer. They are also among the most common types of cancer diagnosed and a leading cause of death in children and adolescents under the age of 20. However, improvements in treatments like those Ellen contributed to have dramatically increased the lifespan of individuals with leukaemia and other types of blood cancer.[11]

Dr Eva Klein: Charting the Margins of the Scientific Map

In the nineteenth century, the average person was more likely to die of pneumonia, influenza, tuberculosis, or even gastrointestinal diseases like diarrhoea than of cancer.[12] Today, cancer is the second leading cause of death worldwide, after cardiovascular diseases. Globally, one in five people develop cancer and one in nine men and one in twelve women die from cancer.[13] The reason for this reversal is that humans are living longer thanks in no small part to scientific understanding of immunity and the development of effective vaccines. Simply put, because many of the fatal diseases of the nineteenth century are now usually no more than inconveniences, humans are living long enough to develop cancer, the risk of which increases with age.[14] It makes sense then that medical researchers would also look to the immune system for the prevention and treatment of cancer.

Immunology – the study of immune systems – is a relatively young scientific field, although observations about immune responses go back to ancient times. Immunisations against smallpox via a technique known as variolation preceded English physician Dr Edward Jenner's late eighteenth-century development of a vaccine for smallpox by hundreds of years. Jenner's story of discovering in 1793 that milkmaids who had been infected with cowpox, a relatively minor disease, were immune to the much more dangerous smallpox is perhaps the most popular version of the origin story of immunology. But it leaves out not only the treatment through variolation of smallpox in China and Turkey long before but also the fact that a woman, English writer Lady Mary Wortley Montagu, contributed to introducing variolation against smallpox in the Western world in the early eighteenth century, before Jenner was even born.[15]

In the late nineteenth century, American bone surgeon William Bradley Coley hypothesised that stimulating a cancer patient's immune system could cure cancer. To test this, he began the first systematic cancer immunotherapy study, treating patients with an injection of toxins that achieved good results. The scientific community was doubtful, however, and his work was not given serious consideration.[16] Research in cancer immunology and immunotherapy continued, existing mainly in the margins over the next half century. It was there that Dr Eva Klein and her husband Dr Georg Klein found cancer immunology in the late 1940s. The Kleins, Holocaust survivors who escaped Soviet-controlled Hungary after the Second World War, worked in the cell research laboratory at the Karolinska Institute in Sweden, where they both studied for and received their medical degrees. As medical researchers in the late 1940s and early 1950s, they initially shared the scientific community's uncertainty about cancer immunology. But frustrations in their assigned research encouraged them to chart their own path, leading them straight into the very field they viewed with scepticism.[17] Individually and together, Eva and Georg's scientific research helped to establish the foundation on which modern cancer immunology stands today.[18]

During her long career, Eva's determination to doggedly pursue difficult research questions resulted in important discoveries. Even before she had completed her medical degree, she refined the process of converting solid tumours into ascites (liquid) form, which made it possible for researchers to analyse tumour strains for understanding of cancer formation and treatment. In the 1970s, she and two colleagues identified natural killer (NK) cells, a type of immune cell that finds and kills cancer cells.[19] Today, NK cells are used in advanced cancer treatments, and the significance of these findings has expanded beyond cancer to viruses, including COVID-19.[20] Her research also provided important new immunological insight into Burkitt lymphoma, an aggressive form of cancer, and Epstein-Barr virus, a virus common in humans that can cause other diseases including various forms of cancer.

Eva's research and that of other cancer immunology pioneers, including Georg, continued to face scepticism through the mid-1980s, when it began to become more accepted. Throughout this period, Eva led the field down paths that few wanted to travel. By the mid-1990s, cancer immunology finally entered the mainstream of science.[21] Medical researchers have been working on developing immunotherapy treatments for cancer and vaccinations that could prevent cancer in the same way as has been done with other diseases that now represent little threat to otherwise healthy people, such as pneumonia, influenza, tuberculosis, and gastrointestinal diseases. Already, vaccines are available for human papillomavirus (HPV), which can cause cervical cancer and head and neck cancer, and hepatitis B virus (HBV), which can cause liver cancer.[22] Some patients with cancer are also being treated with vaccines created from their own tumours as a form of immunotherapy.[23] In time, immunotherapy could replace chemotherapy and radiation, which are highly toxic and can cause other health issues and complications.

Although she worked in the margins for many years, Eva Klein's enormous importance to the field of cancer immunology is now impossible to ignore.

Halina Czajkowska: The Power of Poison

As prevalent as chemotherapy – the treatment of diseases using chemicals – is today, it may seem hard to believe that it has only been used as a cancer treatment on humans since the 1940s. Even more surprising is that the origin of chemotherapy can be traced back to the devastation caused by widespread chemical warfare during the First World War. Among the chemical agents used in the conflict was mustard gas, a compound first deployed by the Germans in July 1917. As a chemical weapon, mustard gas was meant to disable rather than kill, though it did both. Newspapers around the world were abuzz about the devastating effects of this new gas, including painful skin blisters, temporary blindness, and slow death if inhaled. What was unknown at the time was that humans exposed to mustard gas have an increased risk of developing cancer later in life. In the meantime, scientific research into mustard gas began during the First World War and continued into the Second World War, resulting in the discovery that exposure to mustard gas damaged bone marrow and white blood cells. This meant that mustard gas, or some form of it, was a possible treatment for certain types of cancer.[24] In 1942, clinical trials began of nitrogen mustard – a derivative of mustard gas – treatment on a patient with lymphoma. Although other drugs had been used to treat cancer previously, this was the first application of cytotoxic chemotherapy as a cancer treatment.[25]

Cancer occurs when cells grow abnormally and spread to other parts of the body. Chemotherapy utilises controlled amounts of poisonous substances, like nitrogen mustard, to damage or kill these abnormal cells. When the results of the early clinical trials of nitrogen mustard were published after the end of the Second World War, scientists began searching for other substances that could be used to treat cancer through chemotherapy. At this critical moment in medical history, Polish survivor of Nazi persecution Halina Czajkowska completed her studies in chemical engineering and began her career in the world-class cell research lab of Georg and Eva Klein at Karolinska Institute in Sweden. It was there that she learned to analyse white blood cell counts to assess chemotherapy drugs.[26] Halina's skill with this valuable technique and her desire not to lose it after moving to Canada and joining the Collip Medical Research Laboratory at the University of Western Ontario in 1951 led to her part in the discovery of another chemotherapy treatment, vinblastine.

Like mustard gas derivatives, vinblastine is derived from a poison; albeit poison from a more pleasant source – the leaves of Madagascar periwinkle, which contain the chemicals known as vinca alkaloids. As with cancer treatments derived from mustard gas, vinblastine and other cancer treatments derived from vinca alkaloids are especially effective in treating lymphomas and leukaemias. Since the first clinical trials of vinblastine as chemotherapy in 1959, other vinca derivatives have been proven as effective for treating cancer. As had happened following the successful use of nitrogen mustard as a chemotherapy treatment for cancer, the success of vinblastine and its close cousin, vincristine, stimulated interest in the search for other, similarly effective chemotherapy drugs.[27]

Today, vinca derivatives are considered 'the earliest and most durable of the effective anti-cancer agents'.[28] One way of understanding the continued importance of vinca derivatives in treating cancer as well as the increasing prevalence of cancer is through the shortage of vinblastine. Five hundred kilograms of dried Madagascar periwinkle leaves are needed to produce one gram of vinblastine and two thousand kilograms to produce one gram of vincristine.[29] The problem is that the natural supply of this plant cannot keep up with the demand. Children are among the patients most endangered by the shortage, as vinblastine is currently part of one of the most common and effective treatments for paediatric Hodgkin lymphoma.[30] In response to the crisis, scientists are developing new ways of synthetically engineering vinblastine and other vinca alkaloids, which are now at least partly or totally synthetic.[31]

The use of vinca derivatives for chemotherapy has grown exponentially since the 1960s. Although vinblastine's origins are attributed to Canadian researchers Drs Robert Noble and Charles Beer, the reality is that the discovery of the vinca alkaloid's potential as a cancer treatment can be traced back to a young lab technologist who was determined not to let her hard-earned skills go to waste. That she was a woman is almost certainly at least part of the reason why her name is not well known today. But had it not been for Halina Czajkowska, Madagascar periwinkle might just be another pretty flower with poisonous properties rather than the basis of one of the most important chemotherapy drugs available today.

Dr Rita Levi-Montalcini: Vast, Unexplored Areas of the Nervous System

Ancient Greek philosopher and physician Hippocrates (c. 460–c. 370 BC) is often called the 'father of medicine'. Among his contributions was the theory that four 'humors' – blood, phlegm, yellow bile, and black bile – regulated human health and behaviour. As with many ideas of past and present, Hippocrates' humoral theory was neither new nor original. Similar theories are believed to have existed long before Hippocrates' time in South Asia, Egypt and Mesopotamia.[32] The four humors were preceded by at least several hundred years by, for instance, the similarly understood three *doshas* of the South Asian practice of ayurveda.[33] Nonetheless, humoral theory was the guiding principle of most medical practice in Europe and the Middle East for the next two millennia. In the sixteenth century, the first challenges were made to the theory based on scientific study. But it took another three centuries before medical science finally dispelled these deeply embedded beliefs.[34]

The example of the four humors demonstrates how difficult it can be to change scientific understandings and beliefs, even in the face of strong new evidence. While scepticism is an essential component of the scientific process, it also has the potential to hinder the acceptance of new advances for the wrong reasons, not least when gender is involved.[35] Neurobiologist Dr Rita Levi-Montalcini knew this all too well.

Rita never married or had children. But in 1954, a christening of sorts took place for what she later described as her child.[36] That year, in St Louis, Missouri, Nerve Growth Factor (or NGF) was formally named. The gestation of this scientific baby had been long, at least four years. Proper recognition for the importance of its birth took even longer. Over the next several decades, Rita's role in the discovery of NGF was doubted, questioned, and even forgotten. 'For so many years there was doubt that I was – as I really was – responsible for the discovery,' she said in a 1988 interview. 'For a long time people didn't mention how NGF was discovered. My name was entirely left out of the literature. People repeated my experiment and didn't mention my name!'[37]

As frustrating as this would have been, Rita was not one to let gender or any other kind of bias stop her from pursuing her innovative research.

She had not only survived the Second World War as a Jew in Fascist Italy, but she had also defied the racial policies restricting Jews from working in any profession by practising medicine, conducting her research, and even publishing her findings in scientific journals. It was this wartime scientific work that brought her to St Louis in the first place, where she joined the Department of Zoology at Washington University in 1947 at the age of 38.

In these vastly improved circumstances, Rita saw something unusual in her microscope that intrigued her: nerve cells that 'were acquiring a personality not usually attributed to them'. These early findings, she reflected in her memoirs, 'revealed to me the existence of vast unexplored areas of the developing nervous system'.[38] Although her findings went against received knowledge of the time, she pursued them nonetheless. The result was that she identified NGF, the first of what are now called neurotrophins – proteins produced in the nervous system of humans and other vertebrates that regulate the growth, development, and function of neurons, which are nerve cells that control almost every aspect of bodily function, from breathing to moving to thinking.

It took years of research, including a fruitful collaboration with biochemist Dr Stanley Cohen, to identify and isolate NGF and to begin to understand the ramifications of the discovery. In 1986, thirty-two years after NGF was named, Rita and Cohen – who subsequently discovered another neurotrophin, Epidermal Growth Factor (EGF) – were jointly awarded the Nobel Prize in Physiology or Medicine. She was the fourth woman to be awarded this honour. Even then, the importance of the discovery of NGF was still only beginning to be understood. Today, scientists continue to work out the full importance of NGF, which is currently understood to have a role in the prevention of Alzheimer's disease, the modulation of pain, depression, and anxiety, and the regulation of immune functions and diabetes.[39]

Scientific discoveries like Rita's were what helped refute millennia-old beliefs like humoral theory. But being the first to challenge any form of what Rita called 'the generally accepted dogmas' – as NGF did – is sometimes met with a certain amount of anger, fear, and resistance. Certainly, she could have turned back. Instead, she fought such impulses – just as she had done in Italy during the Second World War – and lived to see herself described as 'one of the chief protagonists of the most glorious episode in the modern history of neuroscience'.[40] Rita was truly the 'mother of NGF'.

Dr Halina Neujahr: Converting Human Waste into a Necessity of Human Life

Vitamin B12 is vital for normal functioning of the human brain and nervous system. Also known as cobalamin, B12 contributes to the production of DNA and red blood cells. It also helps in the prevention of heart disease, stroke, osteoporosis, and dementia. But our bodies do not produce this essential vitamin naturally. Humans rely primarily on animal sources, such as beef, poultry, fish, eggs, cheese, yogurt, milk, etc., to achieve the necessary amount of B12 to keep us healthy. Even then, these sources are not enough for adults over age 50 and, especially, individuals with pernicious anaemia – a rare and potentially fatal condition in which the body cannot absorb B12 properly. To offset these deficiencies, B12 supplements are essential, whether in the form of pills or, in the case of pernicious anaemia, injections.

Today, it is relatively easy to handle common and severe deficiencies of B12. But, as with most health issues we take for granted, this was not always the case. It was only a century ago that scientists discovered B12 and its important role in human health. In 1926, medical researchers at Harvard University discovered that pernicious anaemia, which had no cure, could be treated if patients consumed half a pound of raw or lightly cooked liver each day. Soon after, a concentrated liver extract was produced that could be administered daily by injection. Twenty years later, in 1948, scientists discovered that the reason liver treatments were so effective was because liver contained high levels of cobalamin, or vitamin B12. At that time, liver was the only known source from which B12 could be effectively extracted, and liver injections were the only treatment for pernicious anaemia. To make newer, better, and more cost-effective treatments for this deadly disease, scientific researchers were hard at work investigating this new vitamin, including by identifying new sources that could be used for commercial production of B12.[41]

Among those occupied with this research in the early 1950s was Polish Holocaust survivor Halina Neujahr, who was investigating the formation of B12 in sewage sludge – a byproduct of wastewater treatment – as a chemical engineering student at Sweden's prestigious Royal Institute of Technology (*Kungliga Tekniska högskolan*, or KTH). As Halina later explained, B12 – the most chemically complex of all vitamins – was not well understood at the time and its crystals had only recently been isolated from the liver. Before the chemical structure of B12 had been determined by English chemist Dorothy Crowfoot-Hodgkin in 1956, for which she won the Nobel Prize in Chemistry in 1964, Halina became one of the first (if not *the* first) to produce B12 crystals from sewage sludge.[42] In 1953, 29-year-old Halina completed her civil engineering degree in chemistry as the only woman in a class of thirty at KTH. Her research established that B12 could be produced in large quantities in the anaerobic digestion (a form of treatment) of municipal sewage sludge.[43]

Despite the significance of her findings, Halina had no intention of pursuing further education in chemical engineering. At the time, it was unusual in Sweden even for men to pursue advanced degrees in engineering, not to mention women. In any case, Halina had secured not only Swedish citizenship around the time that she earned her degree but also a well-paying job as a laboratory manager at a Swedish candle company. But her professor, Swedish chemist Harry Lundin, had other ideas. From among all his students, he encouraged Halina to continue her research on how B12 formed in wastewater. Halina ultimately decided to give up the job at the candle factory to stay on at KTH, with her salary paid through one of the first scholarships for postgraduate education granted by the Swedish Technical Research Council. Still without any intention of obtaining a doctoral degree, Halina nonetheless took her research down new paths by mapping the organisms in wastewater that were responsible for the synthesis of B12. With Lundin encouraging her to continue, Halina stayed on at KTH and in 1960 became the first woman in Sweden to earn a Doctor of Technology degree in chemistry, and only the third woman in Sweden to earn a Doctor of Technology degree in any specialisation.[44]

Halina's research contributed to knowledge of how B12 and related substances are produced during the processing of sewage sludge. In addition, because sewage sludge is available in large quantities and is a biological hazard that requires sustainable processing, her research also contributed to establishing sewage sludge as a cost-effective way of commercially producing

B12.[45] Based on this understanding, scientists and entrepreneurs have been able to devise ways of producing food for chickens and pigs that would increase the amount of B12 passed on to humans through consumption, thus, offsetting B12 deficiencies. The importance of scientific research on B12 in the twentieth century is evidenced by the fact that three scientists were awarded the Nobel Prize for their contributions to the study of B12. Although Halina's later research on B12 took other directions, her early research established her as a pioneer in biochemical research of B12, a necessity of human life.

Dr Ruth Pick: Your Heart in Her Hands

Practically from the moment 35-year-old Holocaust survivor Dr Ruth Pick arrived in the United States as an immigrant from Soviet-controlled Czechoslovakia in 1949, she devoted her life to cardiovascular science. Cardiovascular diseases (CVDs) like heart disease, heart attacks, and strokes are the world's leading cause of death, claiming nearly nineteen million lives annually.[46] As one of the pioneering medical researchers associated with the Cardiovascular Institute at the Michael Reese Hospital and Medical Center in Chicago, Illinois, Rita made important discoveries about treating and preventing CVDs that have continued relevance today.

Ruth was an unpaid research fellow when she began working at the Cardiovascular Institute in 1949. Thirty years later, she ended her career as the Chief of the Atherosclerosis Research Laboratory of the Institute.[47] In the interim, she published dozens of scientific papers and frequently appeared in the pages of US newspapers. 'Your Heart is in Their Work: The Doctors Pick Are Members of a Team Battling our No. 1 Killer,' read the headline of the *Chicago Tribune* in May 1959, referring to Ruth and her husband, Dr Alfred Pick.[48] Some of the newspaper articles demonstrate the seeming novelty of a woman researcher while others show genuine respect. This is reflected in two Illinois newspaper articles published within weeks of each other in 1962. One headline read 'Woman Doctor Plans Speech', while the lead paragraph of a similar announcement in another newspaper began with 'Noted Heart Fund research investigator, Ruth Pick, M.D. of Michael Reese Hospital and Medical Center, Chicago, will speak to local Heart Association members…'[49] One thing that is clear in all the articles is that Ruth was a respected medical researcher making important contributions to what were, even then, the deadliest diseases in the United States.

Not surprisingly, many newspaper articles referred to Ruth's scientific findings. 'Aspirin May Halt Strokes,' *The Atlanta Constitution* informed readers in May 1979, noting that 'Dr. Pick's work – published in the Journal of Clinical Investigation – is the first to hint daily doses of aspirin can prevent the initial onset of problems.' Ruth was very proud of this discovery.[50] Her findings are still relevant today, even as advances in science and medicine since then have greatly improved understanding of factors like the efficacy and risks of aspirin for different people.[51] Sadly, Ruth's contribution to this area of medical research has been mostly forgotten. Likewise, her contributions to research on atherosclerosis – the hardening of the arteries caused by fat and cholesterol buildup that can lead to strokes, heart attacks, and blood clots – not least her discovery that oestrogen has a preventative effect on the disease, are now mostly forgotten. Ruth acknowledged this reality without any sign of bitterness in the 1990s.[52] Like most scientists, she understood that her achievements were building blocks rather than a final structure.

Of course, with the passage of time and advancements in research, individual researchers tend to be forgotten.[53] But continued gender bias in science is also to blame.[54] An example of this may be found in a 2002 book about scientific advancements in research on cardiac disease in the twentieth century.[55] Of the sixteen physicians whose interviews are featured in the book, only one is a woman, paediatric cardiologist Dr Mary Allen Engle. One of the men interviewed for the book was cardiologist Dr Jeremiah Stamler (1919–2022), who worked with Ruth at the Cardiovascular Institute at Michael Reese Hospital. In response to the interviewer's question about 'outstanding women at Michael Reese', Stamler recalls Ruth not as an important researcher in her own right but as 'an important contributor to some of the work *I was pursuing* on atherosclerosis'.[56] Recent studies show that women's scientific contributions continue to be unappreciated or ignored.[57] This applies to women in STEM now and in the past.

Exacerbating this, a criticism of research on CVDs is that 'most of the investigators in the field were male' and 'most of the early landmark studies in the management of coronary artery disease and its risk factors were restricted to male participants'.[58] In other words, women have been a minority in this area of medical research, as both researchers and subjects. The problematic implications are not limited to the disregard of the important contributions of researchers like Ruth. They also extend to the marginalisation of women's specific health needs and, beyond that, to the specific health needs of a variety of other medically and scientifically underrepresented groups, such as women of colour and transgender women and men.[59]

This is one of the reasons why Ruth Pick's story and those of the other women in this book are important. Representation matters, in science and history. Their stories are examples of the difference women make in STEM. At the same time, they are stories of survival, not just of gender discrimination in 'a man's world' but also of racial and ethnic discrimination in an age of genocide.

Chapter One

Discoveries and Dismissals

> Perseverance and patience are the main ingredients to carry out research. An optimistic outlook is of great help so as not to be easily subdued by disappointment. Without it one is bound to give up in frustration at the inevitable set-backs.
>
> Ruth Pick, MD, 1979[1]

The 1950s was a decade of both progress and regression. In the wake of the Second World War, medical and scientific advances were being made at a rapid pace. Researchers in STEM fields were in high demand and universities around the world expanded their programmes accordingly. Governments increased funding for STEM-related education and research. Progress was the byword of the decade.

The flip side of this progress was regression. Most alarmingly, regression from postwar peace to the Cold War, Korean War, and the nuclear arms race. Less noticeably to most people was the regression from the boom in opportunities for women during the war to hyper-domesticity afterwards, particularly in North America. With men back from war, women were forced to give up the jobs they had filled during the conflict and return to the home. Among the jobs women had to relinquish to men were many in STEM fields. In other parts of the world, this regression was due not so much to a cultural notion of women's place being in the home but to financial and political instability and the demands of postwar reconstruction.[2]

The women in this book were affected by both types of regression. In their home countries before the war, women's education and employment in STEM fields was relatively more common than it was in the countries they eventually settled in after the war. Those who had left or simply not returned to their home countries after the war did so not because of gender bias pursuing an education and career in science but because of factors such as destruction and political uncertainty. Seeking better lives elsewhere, they settled in the United States, Canada, and Sweden, with the intention of continuing the educations and careers that they had begun before and during the Second World War and the Holocaust. It came as a shock to many of them when they encountered the gender bias and gendered culture norms in their adopted countries that contributed to the low participation and success of women in STEM.

Rather than starting chronologically, with the six women's early lives, this chapter takes the 1950s as a starting point because it was a critical juncture in their lives and careers. Studying for or working in their respective careers outside of their countries of origin, they were strongly influenced by their recent pasts. For them, the past and the future converged during the 1950s, setting the course of their careers in the decades to come. The chapter begins with a visit to the lab of Rita Levi-Montalcini at Washington University in St Louis, Missouri, where she made the scientific discovery that ultimately led to her winning the Nobel Prize in 1986. Rita's scientific work in Italy had been interrupted during the Second World War due to Fascist racial laws prohibiting Jews from working at Italian universities. But, in the 1950s, Rita's long and eventful career as a neurobiologist was just beginning to take off.

Progressing through the decade, we visit the lab of cancer researchers Eva and Georg Klein – Hungarian Jews who barely escaped the fall of the Iron Curtain in Hungary – at Karolinska Institute in Stockholm, Sweden. Among those who worked in their lab was Halina Czajkowska, a Polish former political prisoner of several Nazi concentration camps who would later play an important role in a cure for childhood leukaemia while working in Canada. The three other women at the heart of this book – Ellen Loeb, a German Holocaust survivor living in Dallas, Texas, Ruth Pick, a Czech Holocaust survivor in Chicago, Illinois, and Halina Neujahr, a Polish Holocaust survivor in Stockholm, Sweden – are also introduced through their scientific research and discoveries and the associated challenges they faced as women and foreigners studying science and medicine in the countries in which they settled.

Science in Defiance of Danger and Ambivalence

In late 1950, in the Department of Zoology at Washington University in St Louis, Missouri, 41-year-old Dr Rita Levi-Montalcini was revelling in the first glimmers of what became known as Nerve Growth Factor (NGF). Unexpectedly, her former professor and colleague from Italy, anatomist Dr Giuseppe Levi (no relation to Rita; 1872–1965), visited her in her laboratory. With the excitement of a child seeking a parent's approval, Rita invited him to look through her microscope to see her recent discovery. But instead of sharing her joy, the elderly professor responded with what Rita described as rage. Accusing her of amateur mistakes, Levi insisted that she would ruin both her reputation and his if she shared her findings with the scientific community. Rita responded by using the art of distraction. She suggested to Levi that he take a trip to the Grand Canyon for a few days. On his brief return to St Louis, Rita showed him around the city before seeing him off at the airport. Meanwhile, undaunted, she continued to pursue her research. Trusting her instincts, she knew she had found something important and was determined to prove to Levi and any others who might doubt her.[3]

The following year, Rita presented her findings at the New York Academy of Sciences, one of America's oldest and most prestigious scientific societies, and published her results in 1952 in the Academy's journal.[4] She later characterised the dominant response to these presentations of her findings as scepticism and perplexity. 'The discovery of Nerve Growth Factor (NGF) could in no way have been foreseen, nor was it in tune with the concepts that prevailed at the time,' she explained.[5] Rita's response to this dismissal was to take a trip of her own. In September 1952, she arrived in Rio de Janeiro, Brazil to conduct specialised analyses in the laboratory of biologist Dr Hertha Meyer (1902–1990). Like Rita, Meyer was Jewish and had survived the Second World War and the Holocaust. She had managed this by first fleeing Nazi Germany and settling in Italy, where she worked with Levi and trained Rita in various scientific techniques. Then, in 1939, she fled to Brazil to escape the Italian racial laws.[6]

Rita spent several months working in Meyer's laboratory at the Institute of Biophysics in Rio, where – as she later explained – she conducted a 'series of experiments that determined the direction of all my future research' and launched 'an entirely new sector of study in developmental biology by revealing the capacity of certain protein molecules – known as growth factors'.[7] In Rio, she had found the further evidence she needed to convince her sceptics of the importance of her findings. Or so she thought. Either way, her work was far from complete. She needed help to isolate the growth factor she had discovered. This need was fulfilled by biochemist Dr Stanley Cohen (1922–2020). From 1953 to 1959, Rita and Cohen had a remarkable and fruitful collaboration, each bringing their respective skills to research on NGF.[8]

However beneficial the addition of Cohen was to the NGF research, it did not lead to a decrease in the scepticism of the scientific community, at least not right away. The primary issues facing the two scientists were, first, that Rita's original and their joint discoveries were contrary to the deeply ingrained understandings of the time and, second, that she and Cohen were pioneers of an entirely new area of study. When Cohen joined the research, he and Rita began to present findings that seemed to some even more impossible than what Rita had previously presented on her own. As pharmacologist Dr Alice Abbott explains, 'scepticism only increased when Cohen and Levi-Montalcini proposed that snake venom and extracts of mouse salivary glands, both of which also promoted profuse nerve growth, were abundant sources of the factor they were seeking.' This, she says, 'required too great a leap of the imagination to believe'. Worse, according to a prominent biochemist, 'people just didn't believe this was serious biology.'[9]

The doubts of the scientific community may have been based on factors like lack of imagination, fear of change, and so forth, but gender was also a factor. In the United States, the attitude towards women's place in STEM was conflicted, one might even say tortured, during the 1950s and 1960s. The country entered the Second World War in late 1941 in a state of economic depression. By the end of the war, the US was experiencing an economic, scientific, and technological boom. Historian of science Margaret W. Rossiter writes that during this period, women received 'ambivalent encouragement' to enter the workforce in scientific and technological fields, where there was tremendous demand. Women were needed and even wanted in science and technology, but mainly in what were considered low-level jobs supposedly suited for women's competencies, such as science teachers and laboratory technicians, and in which they were virtually invisible.[10]

The 'top' positions were largely reserved for men, thanks in no small part to social attitudes and institutional regulations that negatively affected women, such as the belief that women were unsuited for certain STEM fields and anti-nepotism rules that kept married women scientists from working at the same institutions as their husbands.[11] The result was that relatively few women successfully jumped all the hurdles necessary to earn advanced degrees and find and keep suitable – not to mention high-level – employment in male-dominated fields like the biological sciences, chemistry, and physics.[12] Rita and other women like her were therefore something of an anomaly that at times beggared the belief of the scientific community, not to mention the press.

The newspaper coverage Rita and Cohen received in the 1950s and 1960s reflects the uncertainty felt about women in STEM, particularly those who defied the expectations of the time. Throughout the 1950s, very little had been written in American newspapers about Rita and her findings, with or without Cohen. Typically, what did appear were short blurbs about Rita's presence at medical conferences and, starting in 1959, a few short articles about her joint findings with Cohen, undoubtedly at least partly driven by the attention-grabbing headlines they made possible. 'Ingredient in Mice Saliva Aids Control Nerve Growth', read one Indiana newspaper headline, while another in a Washington, DC newspaper read, 'Snake Venom Promotes Nerve Growth'.[13]

This increase in newspaper coverage coincided with two major changes that occurred in Rita's career in 1958 and 1959: she was promoted to Professor of Zoology at Washington University and Cohen was forced to leave Washington University due to budget cuts. Rita's promotion in 1958 was groundbreaking. She was the first woman promoted to full professor in the natural sciences at Washington University and perhaps the only female Professor of Zoology in the US at that time.[14] Strikingly, this milestone passed virtually unnoticed in the press, even in the local St Louis newspapers. Stanley Cohen, meanwhile, featured in newspaper articles for his continued research at Vanderbilt University in Nashville, Tennessee, where he began working in 1959. A highlight of his career must have been when *The Nashville Tennessean* ran a multiple-page feature on him in its Sunday *Magazine* in March 1962. Although Cohen's research was a continuation of what he began with Rita, she is not mentioned in the article. 'Stanley Cohen is a man of mysteries, a biochemist on a strange hot trail,' enthused the journalist.[15] In contrast, Rita continued to appear in short articles, usually in St Louis newspapers and with bland headlines such as 'Dr. Levi-Montalcini To Receive Award' that almost always named Stanley Cohen as her co-researcher.[16]

The notion that Rita's research was tied to Cohen's, but his research was his own was not limited only to the press. Although both Rita and Cohen were gradually rising in prestige in the scientific community, scepticism about NGF began to diminish only after Cohen discovered another growth factor – Epidermal Growth Factor (EGF).[17] The initial discovery was made while he and Rita were still working together at Washington University but continued after his move to Vanderbilt University, where he first published his findings.[18]

Throughout what Rita later called 'The Saga of the Nerve Growth Factor', she was conscious of the dismissals of and doubts about her and her research.[19] So much so that she briefly set aside her work on NGF in the 1960s before realising, 'I could not give up what was my child, the NGF.'[20] While there is no doubt Rita was angry and frustrated by the response to her and NGF, she also had an incredible capacity to rise above and even seemingly glide over it at times. Her memoir, *In Praise of Imperfection*, reminds us that this ability to persist and even thrive through adversity was born of a mixture of experience and passion. Writing about how she managed to continue her research in Italy during the Second World War despite Italian racial laws prohibiting her from doing so, during Allied bombings of Italian cities, and as the Nazis advanced into Italy, Rita explained: 'The answer lies in the desperate and partially unconscious desire of human beings to ignore what is happening in situations where full awareness might lead one to self-destruction.'[21]

Rita's experiences during the war had taught her to block out as much as possible the danger that surrounded her, not merely because it threatened her life, but also – and perhaps more importantly – because it threatened the pleasure and joy she experienced in her work and losing that would lead to the worst kind of destruction: self-destruction. Her memoirs demonstrate this most clearly in the way she reflected in later life about her work under extreme adversity. Of her tiny, improvised bedroom-laboratory in Turin, where she conducted clandestine experiments with makeshift equipment, she writes in her memoirs of 'the pleasure of managing to work under prohibitive conditions'.[22] Similarly, after escaping the dangers of Allied bombings in Turin and moving to a small country house where she set up another primitive laboratory, her memories revolved around how she began to investigate what 'was to remain the main object of my research in the years that followed'. Rather than inhibiting her research, she wrote that her work 'came into focus and grew in that country milieu probably much better than it would have in an academic institution'.[23] These are the memories of a woman whose work was the love of her life. When faced with adversity, she cloaked herself in her research in an attempt to ignore external threats to her personal and professional existence.

The challenges Rita Levi-Montalcini faced as a Jewish researcher during the Second World War and as a female researcher in an ambivalent America after the war were not the same. But they produced a similar result, each driving her closer to her scientific work and the joy she found there. In her laboratory, Rita found peace and purpose. Under her microscope, she saw beauty and mystery. She discovered things others could not or did not want to see. As best she could, she ignored the dismissals and doubts as she had done with the dangers during the war, until – eventually – the scientific community caught up with her.

'And It Was Not a Sampling Error!'

During Dr Ruth Pick's early years working at Michael Reese Hospital and Medical Center in Chicago, Illinois, women were heartily encouraged to apply… for the hospital's nursing school. In 1949, when 35-year-old Ruth started at Michael Reese as a technician, advertisements ran in northwestern United States newspapers offering scholarships to 'girls of high personal and scholastic standing' for the Michael Reese Hospital School of Nursing. Persuading young women to apply, the ad read, 'As a graduate nurse you are a skilled professional person with wide opportunities anywhere in the world. You will find ready employment in many interesting fields.'[24] Ruth's experiences as a physician and nurse in the Theresienstadt Ghetto and

Prague during and after the Second World War and the Holocaust certainly bore this out, albeit in a macabre sort of way. But while she had found 'ready employment' at Michael Reese as a doctor, the fact that she was initially an unpaid lab technician indicates that while female nurses were in demand, the same was not necessarily true of female doctors.

In November 1948, *The Saturday Evening Post* – a publication with a readership of around four million in the United States[25] – published a feature article titled 'Do Women Make Good Doctors?' which informed readers that any woman considering a career in medicine 'will find enough obstacles on the road to medical proficiency and prestige to remind her that the battle of the sexes, where careers are concerned, is still on'.[26] Women's representation in the US medical profession had barely changed since the 1920s, when women accounted for around 4 to 5 per cent of practising physicians.[27] As with other professions, women working in medicine made significantly less than their male counterparts, a fact one 1951 report insisted must be due to factors such as location (urban versus rural), type of employment (salaried versus independent), and specialisation (the suggestion being that women worked in less lucrative specialisations like paediatrics).[28] In Ruth's case, there was a grain of truth in the last factor since she had no specialisation when she began at Michael Reese.[29] Her husband, cardiologist Dr Alfred Pick, on the other hand, was paid to work at Michael Reese from the start. Nonetheless, there is no question that gender bias in the US medical profession was endemic. Women physicians complained especially of 'the difficulty of securing hospital and medical faculty appointments commensurate with their abilities', *The Saturday Evening Post* reported. '"Often an assistant, never a chief" is a common complaint.'[30] These difficulties would have been compounded for Ruth, a newly arrived Czech-Jewish immigrant and Holocaust survivor.

On balance, although Ruth graduated from medical school at the German University in Prague in 1938 and had practical if rather unconventional experience, she was fortunate to have found a job 'right off the boat', as she put it years later, even if it was initially unpaid.[31] She would soon demonstrate how fortunate both Michael Reese Hospital and the medical profession were to have found her. In the meantime, the Picks were focused on settling into their new lives after three years of waiting to secure the US visas that enabled them to leave Czechoslovakia. The previous year, the metaphorical

Iron Curtain had descended as the Soviet Union sought to isolate itself and the countries under its political control – including Czechoslovakia, Hungary, and Poland, among others – from Western Europe and the United States. The couple arrived in the US by ship in January 1949, just as Soviet leader Joseph Stalin consolidated the Soviet Bloc by establishing the Council for Mutual Economic Assistance, a Communist economic union. Their emigration from Czechoslovakia was supported with loans from family members and help from the Joint Distribution Committee (JDC), a Jewish relief organisation. In the US, they had the support of Alfred's former colleague from Prague, Dr Richard Langendorf (1908–1987), a cardiologist at the Michael Reese Cardiovascular Institute, who helped to secure the jobs for Ruth and Alfred at the Institute.

Like the Picks, Langendorf was Jewish and from Czechoslovakia, where he completed medical school at the German University in 1932, alongside Alfred Pick. Langendorf and Alfred had worked together at the Charles University Department of Internal Medicine in Prague until rising antisemitism and the Nazi annexation of Czechoslovakia in 1939 led to the exclusion of Jewish academics and doctors from working at universities and from practising medicine except in Jewish hospitals. Langendorf managed to escape soon after and found work at Michael Reese, where a committee of doctors – including the head of the Cardiovascular Institute, Dr Louis N. Katz (1897–1973) – was established to find positions for Jewish émigré physicians who had been excluded from practising in Nazi Germany and the territories it occupied.[32] Unfortunately for the Picks, they were not able to secure the necessary visas to leave Czechoslovakia. They experienced the full weight of the Holocaust but miraculously survived and reestablished contact with Langendorf after the war.

Reunited at Michael Reese Hospital, Langendorf and the Picks worked in what must have seemed a sanctuary from antisemitism, genocide, and repression. The hospital was established by the Hebrew Relief Association of Chicago with a donation of 200,000 US dollars (around 5 million US dollars today) left in the will of Jewish real estate developer Michael Reese, who died in 1877. One of the conditions made by Reese's heirs was that the hospital, which opened in October 1881, provided care to anyone 'regardless of creed or nationality'.[33] Throughout its history, the hospital honoured this condition, including by serving Chicago's large immigrant and African American populations, regardless of whether the patients could pay or not. The fact that it was a charitable institution did not preclude it from also being a world-class research hospital, with many Jewish doctors among those making significant medical advances.[34]

It did not take long for Ruth to make her mark in this sympathetic but male-dominated environment, where she worked in the pathology laboratory after deciding to specialise in histology, the study of the anatomy of biological tissues, cells, and organs through a microscope.[35] Soon, Katz gave Ruth a paid position for which she made twice as much as her husband while he was doing his medical internship, as Ruth recalled with pride.[36] The Cardiovascular Institute team had recently begun studying the effect of oestrogen on atherosclerosis, a hardening of the arteries that can lead to heart attack, stroke, blood clot, and aneurysm. When the team was on the verge of abandoning their experiments, which were conducted by cutting open and comparing the aortas of two groups of chickens – one fed oestrogen supplements along with a high-cholesterol diet and the other the same diet without oestrogen supplements – Ruth suggested that they examine the coronary arteries instead. It was an innovative idea that went against the established understandings of the time. 'It had always been assumed that what happened in the aorta (the great trunk leading from the heart) automatically happened in the coronary arteries of the heart itself,' explained *Chicago Tribune* journalist Lillian Pompian in 1959. 'And it had always been easier to study the comparatively large aorta than the tiny coronary arteries of chicks.'[37] As a result, Ruth's suggestion was not taken seriously. But she insisted and was eventually permitted to conduct the experiments.

When Ruth brought the team the first evidence that the chick given oestrogen had no sign of atherosclerosis while those that were not fed oestrogen had severe atherosclerosis, her findings were treated with scepticism.[38] One colleague, cardiologist Dr Jeremiah Stamler, later admitted that he ignored the results, assuming Ruth had made a sampling error, a response Ruth remembered years later with exasperation. Her reaction was to 'take those little hearts apart, piece by piece, totally', she said years later with a note of exasperation in her voice. 'And it was *not* a sampling error,' she said with a light slap on her leg for emphasis.[39] Her own determination was bolstered by the fact that she was supported by Katz and another colleague, cardiologist Dr Simon Rodbard (1911–1975), with Katz telling Stamler, 'You've got to listen to what this lady is saying.'[40] In late 1952, Ruth proved her important findings beyond the shadow of a doubt. Pompian, a trailblazing feminist journalist,[41] opened her article in the *Chicago Tribune* with a description of this defining moment:

> On a cold morning of December 1952, dark-eyed, pint-sized (5' 2') Dr. Ruth Pick, staring intently thru [sic] her microscope at specimen after specimen of animal heart muscle, felt a mounting excitement. What she saw might mean an important link in the conquest of the greatest single menace to health in the United States today and the cause of most of those 'heart attacks' which all too often strike in the prime of life.[42]

Pompian was right. Since 1950, deaths caused by cardiovascular diseases have dropped significantly in most of the world.[43] This has been thanks to scientific discoveries and advances like Ruth's. Up to the 1960s, coronary atherosclerosis was a significant cause of heart disease deaths. But this trend shifted in subsequent decades due in part to preventative methods.[44] Ruth's discovery contributed to this shift by increasing understanding of potential causes and treatments of atherosclerosis. Late in her life, Ruth said this was one of two discoveries she made in her career that she felt most proud of, saying:

> And then we found out that estrogens have a protective effect on heart disease, and I am very proud of that because we were one of the first who showed that, even though it was only in chickens. But later on, we proved it in monkeys and then people started doing it with humans.[45]

Ruth's ingenuity and persistence demonstrate how the quote that began this chapter, which she wrote in 1979, was not empty rhetoric. Nor was her significant contribution to the discovery that oestrogen can prevent atherosclerosis a matter of luck or scientific brilliance. Neither of these would have mattered had Ruth given up in the face of scepticism and dismissal.

A Very Important Problem

In the Swedish laboratory at Karolinska Institute that Halina Czajkowska had left in 1951, 29-year-old Holocaust survivor Eva Klein was making her own advances in cancer research. On 17 March 1954, one of Sweden's major national newspapers, *Svenska Dagbladet*, published an article about Eva and her husband, Dr Georg (sometimes spelled George) Klein, that began on the front page and announced, 'Swedish researchers find important key to the problem of cancer.'[46]

Eva had already completed her medical degree and was working on her PhD in tumour biology at Karolinska, while Georg had already completed his degrees and held the position of assistant professor in the Institute's department of cell research. The duo had made incredible progress since joining the department – Georg in 1947 and Eva in 1948 – after escaping Communist repression in Hungary. As foreign medical students with no knowledge of Swedish, both Georg and Eva had been hired immediately by Professor Torbjörn Caspersson (1910–1997), head of the cell research department at Karolinska. In addition to the job, Eva also received something that Ruth Pick and Halina Czajkowska had initially received in short supply in the United States and Canada: unconditional professional and personal respect. On arriving in Sweden, Eva was invited to lunch by Caspersson, and they were joined by a German professor. She communicated with both men in German and, to her surprise, felt that she was treated as an important person.[47] Not surprisingly, Eva bloomed at Karolinska.

The accomplishment highlighted in *Svenska Dagbladet* was Eva and Georg's development of a method of converting solid tumours in mice into ascitic tumours, or what the newspaper called 'pure cultures' of tumours. As the journalist described, the couple had been working for years 'with a new procedure sought after in cancer laboratories around the world, which for the first time enables the pure production *en masse* of growing tumor cells, i.e. diseased cells in liquid form released from healthy ones'. The immediate significance of this was that, 'new paths have been opened for the hitherto complicated study of the tumor's chemical composition, its growth rate and the enigmatic anatomical structure of the diseased cell.'[48] The following year, 1955, Eva completed her PhD with a dissertation on ascites tumours so groundbreaking it is still cited in scientific texts to this day.[49] She was quickly promoted to associate professor at Karolinska.

The couple's original intention had been to stay in Sweden only as long as it took for both to receive their medical degrees. But they had entered an emerging area of research – or 'a very important problem' as Eva later called it – and received help and support from Caspersson and Karolinska.[50] Despite offers elsewhere, the couple remained in Sweden for the rest of their lives. The very important problem that kept them there was cancer immunology – the study of immunological responses to cancerous tumours. When the Kleins became involved with cancer immunology – also called tumour immunology – it was both an emerging field and an established one. As Eva and Georg later wrote in 1995, tumour immunology 'was born like the hero of Kalevala, the Finnish national epos [epic], the great poet Väinemöinen, after an immensely prolonged pregnancy. Väinemöinen was 600 years old at birth. TI [tumour immunology] was not that old but, like V, he was regarded old and wise already in the cradle.'[51] In addition to being a very old infant, the study of cancer immunology was also controversial.

Since the nineteenth century, when physicians first theorised links between the human immune system and incidences of cancer and its treatment, the scientific community's attitudes about these links were generally dismissive.[52] Among the sceptics were Eva and Georg, who later wrote that 'those of us who actually worked with cancer cells were profoundly sceptical' of the 'extravagant' claims made in 1955 by their former Karolinska colleague, Dr Bertil Björklund, that he had developed a cancer vaccine along with two American physicians, Ruth M. and John B. Graham.[53] The Kleins were only partly correct in their assessment since the Grahams are acknowledged as conducting the first cancer vaccine trial – albeit in 1959 and without Björklund – though the results were largely overlooked. By then, Eva and Georg were fully immersed in cancer immunology research.

In 1957, the Department of Tumour Biology was established at Karolinska with Georg appointed as a full professor and department chair. Eva, though also part of the new department, remained an associate professor for another twelve years. Despite the respect Eva received from Caspersson and other researchers, she was still a woman in a male-dominated profession. During the 1950s, women accounted for between 10 and 13 per cent of people with

Doctor of Medicine degrees in Sweden.[54] At Swedish universities, including Karolinska, entrance requirements were the same for men and women, but the figures reveal that women were enrolled in medicine in much smaller numbers. In fall 1950, just 20 per cent of the 864 students enrolled were women.[55] Karolinska's first female professor was appointed in 1937 and the second in 1965. Eva's turn finally came four years later, in 1969.

Although she and Georg conducted research both independently and in collaboration, Georg received most of the invitations to visit, research, and lecture around the world, at least during their early years at Karolinska. Requests for researchers' participation and collaboration could be explicitly gendered. In 1950, a major private research foundation in Sweden asked Caspersson to, as Georg and Eva later put it, 'choose two young *men* for an urgent mission' to visit the United States and report on cancer research being done there.[56] Georg was selected. He was sent to the Institute of Cancer Research at Fox Chase in Philadelphia, where he worked for several months with a former colleague of Caspersson, geneticist Dr Jack Schultz. Whether the selected scientist actually had to be a man is doubtful since the Fox Chase laboratory, under the direction of Dr Stanley P. Reimann, had gained a reputation for hiring women on equal terms as men already in the 1930s.[57] Eva would not have been out of place working among women like biochemists Dr Grace Medes (1886–1967), who was hired in 1931, and Dr Elizabeth Knight Patterson (1874–1966), hired in 1944. But if Eva was disappointed, she took it in stride. After all, the couple had a young son, and Eva was conscious of the times in which she lived and of the reality that she was a woman working in a man's world.[58] At the same time, she was married to a man who wholeheartedly supported her scientific research.

Georg's trip to the United States contributed to an important turning point in both their careers. On his return to Sweden, he brought with him two hundred inbred mice given to him by geneticist Dr Theodore S. Hauschka (1908–1999). Eva and Georg's early cell research for Caspersson – which involved measuring light absorption in living cells – had been unsuccessful and frustrating. The problems were manifold. The number of cells they had access to was limited (they had one rooster to extract cells from), the process of creating tissue cultures for examination was laborious and time-consuming, and their equipment kept killing the cells before they could draw any conclusions. Without improvements in all these areas, they faced failure. Then came two glimmers of hope. The first occurred when Eva had an ingenious idea after a lecture by German cancer researcher Hans Lettré (1908–1971) on the Ehrlich ascites tumour.[59]

Ascites is the abnormal collection of fluids in the abdomen caused by a variety of diseases, including cancer. Ehrlich carcinoma is a type of breast cancer found in mice that grows in solid (tumour) and ascitic (liquid) forms. Because ascites contains many living cells that can be studied without having to create tissue cultures, the Ehrlich ascites tumour is ideal for studying living cells in large numbers. Eva immediately recognised the importance of this for their research. The only problem was that they needed a large number of inbred mice – essential in cancer research for their lack of genetic diversity – that were unavailable in Sweden.[60] The second glimmer of hope for the couple was the colony of mice Georg brought back from the US, which made it possible for them to conduct their groundbreaking immunology research on ascites tumours for years to come. These continuously inbred mice were used by Eva in the research for her groundbreaking medical dissertation.[61] A photo of Halina Czajkowska from around 1950 shows her working with these mice in the Caspersson laboratory. The mice also featured in the 1954 *Svenska Dagbladet* article. By that point, Eva and Georg had around 5,000 mice, from which they were growing thirty-five tumour strains that cancer researchers could use for analysis. In just four years, they went from being supplicants making 'emergency' trips to the US for knowledge and mice to being internationally recognised medical researchers receiving requests for tumour cultures from research labs from around the world.[62]

Eva's inspiration to use Ehrlich ascites tumours in mice and Georg's acquisition of the original colony of inbred mice had set the stage for what one immunologist describes as 'the groundbreaking work on tumor immunology and the origin of tumorigenesis [the formation of cancer], which the Kleins and their colleagues would perform over the following decades'.[63] In other words, Eva and Georg's early research made it possible for researchers to develop better understandings of both how cancer develops and how to prevent and treat cancer. On this foundation, they and other researchers continued to build, and still do. Eva's own best work was yet to come. In time, she would receive invitations as readily as Georg. Not necessarily because of decreased gender bias in science but because she was a recognised scientific force in her own right who could not be ignored simply because she was a woman.

Not Today, Death

In December 1958, doctors in Texas diagnosed 2-year-old David Conner with leukaemia and estimated he had four to six months to live. While the names of these doctors have been lost to history, that of haematologist Dr Ellen Loeb has not. When David was brought to the Wadley Research Institute in Dallas, 38-year-old Ellen was determined not only to prolong his life but also to make it as normal as possible. She was successful in both respects. The treatment and care Ellen provided David with enabled him to live a relatively normal life for nearly two more years. 'David runs and jumps, rolls and tumbles, shouts and sings like other little boys,' reported a Texas newspaper in July 1960.[64] Of course, there were limits to what Ellen and medicine could do for an aggressive disease with no cure. 'His playing is not limited, but he does take quantities of medicine and makes frequent trips to Dallas for blood and bone marrow tests,' reported his hometown newspaper around the same time.[65] In early September 1960, David died just before his 5th birthday.[66]

Ellen's desire to delay the death and improve the quality of life of her patients as much as possible was deeply rooted in her Holocaust experiences. Not just in terms of her own experiences of persecution and survival but also of the experiences and fates of her family and fellow prisoners in Nazi concentration camps and ghettos. 'Her devotion to her work has a very strong link with the past,' a colleague said of her in 1961. 'She clings to each patient even if the case is hopeless.'[67] For Ellen, there was always and only hope. And she had determined early on that giving up was not an option. Of her patients, she once said, 'we can only offer time, but time is hope – and hope is life.'[68]

Ellen, who was originally from Germany, began her medical career as a nurse while imprisoned by the Nazis in Westerbork transit camp in the Netherlands, where she and her parents were sent after their arrest by the Nazi Gestapo on 8 November 1942. At age 22, she lacked any formal medical qualifications. From Westerbork, the Nazis sent the Jewish prisoners either to an immediate death in extermination camps or to concentration and slave labour camps where there was a slim chance of survival.

For a time, Ellen worked as a nurse in the paediatric unit at Westerbork. The children's immediate fate was doubly in the hands of the doctors and nurses who were prisoners themselves. They had to help them recover their health to save their lives both as human beings and as Jews slated for extermination. 'We were to make them well so that they could either go to work or at least be brought to Auschwitz to be done away with,' Ellen wrote to her sister soon after the end of the war.[69] But there was a loophole: a sick child could not be transported. So, when possible, Ellen and her colleagues tried to delay children being transported by feigning illnesses. 'There was one child in particular we could keep from being transported only if he became ill, and that might save his parents as well,' Ellen wrote. 'So we gave him a milk injection. The poor boy developed a high fever, but that saved his family.'[70] The experience frightened Ellen but she felt good about the work she did.

Like Ruth Pick, Ellen was also imprisoned in the Theresienstadt Ghetto, where she continued working as a nurse. While there, she had an unknowing encounter with her future career when was put to work with German unit working with blood diseases.[71] Later, after miraculously surviving the Holocaust, Ellen specialised in haematology and spent her medical career researching and treating patients with leukaemia – a cancer of the blood – and other blood-related diseases, such as haemophilia. In Dallas, Texas, Ellen faced similar struggles to Ruth Pick in Chicago. In 1950, when Ellen was attending Southwestern Medical School of the University of Texas in Dallas and also working as a nurse to support herself, only around 29 per cent of American women were in the workforce, and these were primarily as teachers or in low-skilled jobs such as office and factory workers and retail salespeople.[72] When Ellen graduated in June 1952, as one of just four woman in her class of sixty-two, she joined a select group of women who comprised around 5 per cent of medical doctors in the US.[73]

Women in any male-dominated career in the US in the 1950s had to work both harder and smarter than men to achieve a reasonable level of recognition and respect. Often, even this was not enough. Gender bias was rife, and not just among men. One woman paediatrician in Ohio told a journalist that women physicians should not attempt to 'compete' with men in fields such as surgery because of their 'physical limitations – such as withstanding the rigors of a four or five-hour operation'.[74] In 1962, psychologist Dr George W. Crane featured a 'case' in his syndicated newspaper column 'The Worry Clinic' – which appeared in around 250 US newspapers – about an anonymous 'private secretary in a large medical clinic' who expressed her dislike and mistrust of women in positions of authority, including physicians. 'Neither would I ever consult a woman physician or dentist or lawyer, unless absolutely forced to do so, and I am working right now in a clinic with five male doctors and two female [sic],' she was quoted as saying.[75] Whether these sentiments were those of the woman or the columnist is open to debate.

Ellen had something of an advantage at the Wadley Institute. In Chicago, Ruth Pick had the advantage that she was not the only Jewish immigrant or even the only refugee from Nazism. Similarly, in Dallas, Ellen worked alongside physicians from all over the world, including Yugoslavia, Turkey, and Argentina.[76] This diversity may have helped her feel more comfortable, at least as an immigrant, if not as a woman. As both immigrants and women in STEM, Ellen and the other women featured in this book were at a double and even triple disadvantage. They had to contend not only with gender bias but also bias against foreigners. Not to mention the fact that all but one of them – Halina Czajkowska – was Jewish.

Racial discrimination in parts of North America in the 1950s was not reserved only for Blacks but also affected Jews, who were often restricted – officially and unofficially – from living in certain neighbourhoods, staying in white-only hotels, joining 'gentile' country clubs, and even visiting certain public beaches.[77] This was true in Dallas, home to Texas's second-largest Jewish population, which rose from 10,000 in 1950 to 16,500 in 1960. Jewish residents of Dallas were rarely the recipients of antisemitic verbal or physical violence, as Black residents were, but were still excluded from most social and professional clubs and organisations and public office because they were Jewish.[78] They could also face barriers as professionals. 'No profession or academic discipline reacted with more hostility to the prospect of Jewish refugees from Nazi-dominated Europe immigrating to the U.S. than did the American medical profession,' writes journalism professor Laurel Leff of the 1930s and 1940s.[79] By the 1950s, the situation began improving slightly, but Ellen and others like her – Jews, foreigners, women – still had to overcome multiple biases that could hinder professional opportunities and recognition, especially in male-dominated STEM fields.

Successfully navigating these challenges requires certain personal characteristics and qualities. 'Historically, women have made an impact on hematology not through sheer numbers, but through drive, intelligence, and compassion,' writes haematologist Lanie K. Francis.[80] Ellen may have had to deal with gender and other barriers, but she also had all three of these qualities in abundance, a fact that was clear to others. 'It has been said that her duties call for a difficult balance of compassion with scientific skill, science with shattering sympathy for doomed children and gallant parents,' reported one newspaper about Ellen in 1964.[81] As for her drive and intelligence, Ellen was a passionate researcher always working on the cutting edge of science. She conducted early clinical trials using many new chemotherapy treatments during her career, including vinblastine, the drug Halina Czajkowska had contributed to discovering, and cisplatin.[82] She also gained notoriety for contributing to the development of autologous bone marrow transplant – a method for removing and preserving bone marrow from leukaemia patients and replanting it after chemotherapy treatment to kill malignant cells – a procedure still in use today.[83]

Ellen Loeb's success as a haematologist can be judged based on how well she overcame the obstacles as a foreign, Jewish woman in a male-dominated field. It can also be based on her dedication to contributing to breakthroughs in the treatment of leukaemia and other diseases of the blood. But while these are both certainly a strong measure of her professional success in STEM, her success may also be seen in how one of her patients remembered her in 1996, sixteen years after her untimely death in 1980. In September 1996, a hearing was held before the US House of Representatives regarding the passage of legislation to create a fund that would provide restitution to haemophiliacs who were infected with HIV after receiving contaminated blood during transfusions. Among those who submitted testimony was one of Ellen's former patients, Bob Baldwin. Baldwin began his testimony by recalling Ellen's Holocaust experience, quoting her as saying 'We are strong… we can work in the fields.' He continued, explaining:

> These were the words of my childhood hematologist facing the 'gates of hell' in Nazi Germany. She managed to save herself and her mother 'Ommi' from the gas chambers of [Auschwitz] by saying these words to her Nazi captors as she approached the head of the line.

Baldwin credited Ellen with saving his life, too, by navigating him to a treatment for haemophilia that did not involve the use of HIV contaminated blood. 'We were fortunate to be under the care of dear Dr. Loeb,' he reflected.[84] There can be little doubt that Ellen's Holocaust experience and her career as a STEM pioneer were intimately entangled.

The Beauty of Microorganisms

On Monday, 30 May 1960, 35-year-old Halina Neujahr had what could be described as a Cinderella moment. Elegantly dressed and wearing her custom-made doctor's hat, she dined and danced the night away at a party to celebrate her and Sweden's other newest doctors. Held in the Blue Room at Stockholm City Hall, where the Nobel Prize Banquet is held each year, the party was attended by dignitaries including Prince Bertil, a younger son of the Swedish king. The next day, a photo of Halina dancing with the governor of Stockholm appeared in local newspapers around Sweden, including the front page of the morning newspaper *Stockholms-Tidningen*.[85] Literally overnight, Halina had become a sensation, recognised for her groundbreaking accomplishment of becoming the first woman in Sweden to earn a Doctor of Technology degree in chemistry and only the third woman in Sweden to earn any Doctor of Technology degree. No one, not even Halina, could have imagined that she would find herself in this position.

Thirteen years earlier, a shabbily dressed Halina had appeared at the office of Professor Ragnar Woxén (1905–1994), rector at Sweden's prestigious Royal Institute of Technology (KTH), asking in broken Swedish how she could study chemistry there. With some difficulty, she had managed to get time off from her job at the Swedish postal service, even though she was aware that she had little chance of studying at KTH, which did not admit foreigners as regular students with a path towards a degree. But Woxén was impressed by Halina's grades and suggested to her how she could obtain the further experience she needed to be admitted as an 'extra student' who had the possibility to eventually become a regular student and pursue a degree. One year later, in 1948, Halina met all the requirements and began studying chemical engineering as an extra student at KTH, thus starting on the path that would lead to her Cinderella moment in Stockholm in 1960.[86]

Of course, Cinderella moments are just moments. Halina's transformation from foreigner trying to attend one of Sweden's most prestigious universities to groundbreaking female researcher dancing with dignitaries in Stockholm City Hall was long and difficult. Her journey began years before she appeared at Woxén's office, in the fall of 1939, when 14-year-old Halina's formal education was halted following the Nazi invasion and occupation of Poland.[87] Likewise, her journey did not end with that undoubtedly magical evening in May 1960. Her more than thirty-year career was marked by the realities of being one of the few female biochemists in Sweden. Like the other women in this book, Halina's career as a STEM pioneer was characterised by both discoveries and dismissals.

Halina was making important scientific discoveries already in the early 1950s, while still an undergraduate student at KTH. Like Halina Czajkowska, Halina Neujahr had chosen to study for an engineering degree in Sweden, where the culture was what could be called a young boys' club. As foreigners, both women were free of the cultural belief held by Swedish women at the time that engineering was the domain of young men. 'In chemistry, I was alone,' Halina N. reflected later in life. 'It wasn't that strange, but if I had known it would happen, I might have been discouraged. But I hadn't been indoctrinated to think that it wasn't something for girls.' For her, the biggest challenge of being a student was not dealing with the male culture of heavy drinking and swearing. By her own admission, she rather enjoyed learning and using the new vocabulary. Instead, her greatest struggle was understanding what was being said in the lectures, which were naturally all in Swedish. 'The first year was terrible because I didn't understand what they [the professors] said, I didn't understand anything.' To compensate, she devised strategies for learning the material while also learning the language. On top of it all, she continued to work to support herself. Despite everything, she successfully passed the exams and became a regular student with a path towards earning her degree.[88]

To earn her degree, Halina had to choose a specialisation. With no biotechnology option available at the time, she chose food chemistry and was assigned to study the formation of B12, still in its scientific nascence, in sewage sludge from a wastewater treatment plant in Stockholm. Where most people would likely meet this proposition with disgust, Halina saw something beautiful. 'It was attractive as a target because it [B12] is beautiful red crystals that previously could only be isolated from liver,' she explained. 'I thought, could there be such a beautiful substance in sewage?'[89] To find out, Halina had to struggle not only through the stench and the many steps required to separate the crystals from the sewage sludge but also through her negative associations with sewage.

During her time as a prisoner in the Majdanek concentration camp, Halina was forced to empty the latrines used by all the prisoners, which required two prisoners carrying a heavy container filled with human excrement, which Halina estimated must have weighed around one hundred kilograms, to a designated location and emptying it. On one occasion, Halina stumbled while carrying one of these containers with another prisoner and accidentally spilled some of the contents in the process. As punishment, a female Nazi guard shoved Halina's head into the bucket, smothering her in the excrement. As Halina recalled, it was the 'greatest humiliation I ever experienced'.[90]

Given that Halina's educational goals were simply to study and learn a profession in the process, the handling of sewage sludge seems a brutal ask considering her past experiences. But, of course, only Halina knew of this trauma at the time, and she was loath to tell anyone. Somehow, she managed to turn the horrors of the concentration camps on their head and use them as professional inspiration. That she could invert her traumatic experiences by seeking 'beauty' in sewage is nothing short of remarkable. After eight months of work, she managed what no one else had done before her: 'to get crystals of B12 (0.8 mgl) from hundreds of liters of stinking sewage sludge,' she explained in 1995.[91] Her research on B12 contributed to making KTH 'a world leader in the extraction of vitamin B12, the market's most expensive fine chemical, from sewage' in the 1950s, according to Halina.[92] Even late in her life, she kept a photo of 'these beautiful red crystals' in her office.[93]

Halina's potential as a scientific researcher did not go unnoticed. Although she had not intended to pursue a doctorate degree, her undergraduate supervisor, Professor Harry Lundin, saw something special in Halina and encouraged her to continue her research and pursue her Doctor of Technology degree in chemistry. As good as she felt on that beautiful evening in May 1960, when she celebrated this milestone, it did not take long for her to realise that her elevated position ultimately led to her greatest professional challenges. After working for two years in the United States, she was lured back to KTH by Lundin with the promise of becoming 'acting professor' of biochemistry, with the task of building the new department. But, in 1967, rather than being promoted to a permanent position as head of the department, she was replaced by a male professor. Over the next three decades, male heads of department came and went. Each time the position was open, Halina was asked to fill the gap by serving as acting head of the department. She applied for a professorship and the permanent position of head of department but was denied every time. The reasons were contradictory at best. 'When I applied for biochemistry, it was said that I was actually a microbiologist, when I applied for microbiology that I was actually a biochemist,' she told Swedish governmental investigators in the mid-1990s.[94]

The last time her application was rejected, sometime in the 1980s, Halina felt humiliated by her treatment and formally appealed the decision. In her appeal, she cited 'several glaring contradictions' in the expert committee's decision documents. This, according to Halina, 'caused some headaches' for the KTH lawyers and precipitated the mysterious disappearance of the decision documents she cited. Although her appeal was denied, she was eventually granted a personal professorship (*rådsprofessur*, in Swedish) by the Swedish government in 1989 following a request by KTH. When she retired in 1991, she was given emeritus benefits that allowed her to continue her research in a small laboratory space of her own at KTH.[95]

Halina summarised her long career in STEM by recalling she 'felt completely equal' during her education and never felt 'socially outcast' among her male colleagues during her career. 'I didn't feel bad as a woman or as a member of a minority until I started competing for positions,' she explained. In another interview, she doubled down on this statement when she said that half the time she was passed over for promotion it was because she was 'beaten by more meritorious men'. But the other half could be attributed to the fact that she was a woman of foreign origin who did not fit into the established pattern.[96]

Despite her tortuous path to official recognition, Halina's career continued to result in discoveries that have contributed to improving life and health. Always seeking the 'beauty' in her research, Halina became particularly successful at studying 'difficult' microorganisms. In the 1970s and 1980s, this type of research led her to discover 'completely new enzymes for breaking down phenol and other environmental toxins'. Her findings were internationally recognised and contributed to knowledge of how to eliminate biological hazards from the natural environment using biological systems.[97] In other words, she discovered new substances for eliminating compounds found in wastewater that can cause environmental damage and prevent them from entering water sources in the first place.[98]

It is unlikely that Halina Neujahr and Ruth Pick ever met. Yet Halina exemplifies Ruth's statement quoted at the beginning of this chapter. In addition to having the perseverance and patience necessary to carry out her research, Halina's optimism – epitomised in her ability to find beauty even in sewage – certainly helped her to bear the disappointments and setbacks that came with being a woman and a foreigner in STEM research in Sweden.

Discovering Life in the Flower of Death

Around the time Ruth Pick was revelling in her contribution to the discovery of the link between oestrogen and atherosclerosis, 26-year-old Halina Czajkowska was making a discovery of her own at the Collip Medical Research Laboratory at the University of Western Ontario. Halina had arrived in Canada in April 1951 with hopes of finding a job as a chemical engineer. Her disappointment in that respect became a scientific gain in the long term. Soon after she was hired as a lab technologist in the Collip Laboratory in May 1951, she was set to work immediately by her supervisor, Dr Robert Noble (1910–1990), with the task of helping to determine if periwinkle leaves had any effect on blood sugar levels.[99]

Periwinkle – in Latin, *vinca* – is a low-lying, fast-growing, herbaceous perennial native to Europe, northwest Africa, and southwest Asia, with distinctively pale purple-blue flowers. The colour periwinkle is named after the smaller type, *vinca minor*, which bears a lighter colour than its larger cousin, *vinca major*. Symbolically, the periwinkle represents serenity and love. But in Italy, the plant is called the flower of death (*fiore di morte*) because it was traditionally placed on children's graves.[100] Periwinkle leaves are also poisonous. Madagascar periwinkle is often a vibrant pink as its former Latin name *Vinca rosea* (more recently, *Catharanthus roseus*) suggests. Rather than death, it could be associated with life insofar as its use as a medicinal plant goes back to Mesopotamia circa 2600 BC.[101] This was the *vinca* species Noble gave Halina. Her job was to give lab rats a concentrated oral extract of the leaves and then measure their glucose levels as part of Noble's study to determine if such extracts could replace insulin injections. But Halina had other ideas. She would do the tests Noble assigned but she would also use the skills she had developed in Stockholm, Sweden at the Karolinska Institute.[102]

Halina's time at Karolinska, Sweden's third-oldest medical school and a world-renowned research institution, had been short – barely one year. But she had a solid track record of accomplishing a great deal in little time. When she arrived in Sweden in July 1945 as a very ill 18-year-old liberated concentration camp prisoner, she spoke no Swedish, and her formal education had been interrupted when Nazi and Soviet troops invaded Poland in 1939. In under five years, she finished her upper secondary education at a school for Polish refugees in Sweden – completing two years of education in one year – and earned her degree in chemical engineering at a vocational school in Stockholm called *Stockholms Tekniska Institutet* (STI). Immediately afterwards, she began working in the cell research department at Karolinska alongside Holocaust survivors Georg and Eva Klein. There, according to medical historian and haematologist Jacalyn Duffin, Halina 'mastered a number of delicate biological techniques, including the determination of blood sugar and of white [blood] cell levels' for the assessment of chemotherapy drugs.[103]

Globally, few women pursued degrees in engineering before the 1950s. This was only partly due to the systematic exclusion of women from engineering programmes. In Sweden, restrictions preventing women from enrolling in higher technical education were lifted in 1921. But women's uptake was slow due to the persistent cultural belief that 'this was an education for "young men"'. The result was that by the 1950s, women accounted for just over 5 per cent of the student body at the Royal Institute of Technology (KTH),

Sweden's leading technical institute.[104] Halina had thus done phenomenally well for herself in Sweden. When she left the country, she carried a valuable letter of recommendation from Dr Georg Klein, a rising star at Karolinska. She quickly realised, however, that employers in Ontario were not necessarily inclined to hire a woman chemical engineer.[105]

Although the letter from Georg Klein impressed laboratory boss Dr J. B. Collip (1892–1965) enough to secure her position in his laboratory, it was as a laboratory technologist – a position more open to women in Canada than that of chemical engineer.[106] Women's progress in engineering had gotten off to a more promising start in Canada than it had in Sweden but was arguably worse subsequently. Canada's first woman chemical engineer, Rona Hatt, completed her degree at the University of British Columbia in 1922, the year after Sweden lifted its restrictions on women studying engineering. Her accomplishment hardly opened the floodgates in Canada. It took another twenty-five years before another woman graduated from the University of British Columbia with an engineering degree.[107] In general, the number of women graduating with degrees in engineering in Canada was relatively small well into the 1950s. In 1951, Ontario had 753 chemical engineers, only one of whom was female.[108] Halina would have quickly recognised that she was not likely to become the second.[109]

Despite the disappointment, she later explained that she was 'anxious to preserve my expertise in cell counting, acquired in Sweden but being lost, through nonuse [sic], in Canada'.[110] When she was given her first assignment, she decided to conduct white blood cell counts on the rats in addition to measuring their glucose levels. She did not tell Noble.

The glucose experiments went from bad to worse. The oral extract – a *vinca* infusion or 'tea' – had no effect on the rats' blood sugar. Although the purpose had been to find a substitute for insulin injections, the rats were then injected with the extract. Not only did the injections have no effect on the rats' blood sugar, it killed them. Halina's clandestine white blood cell counts were another story.[111] As Duffin explains:

> When the experiments turned to injections on or before September 1952, she observed the decline in white blood cell counts. At first she thought it was an error, but the finding occurred on repetition, and she accumulated and

> confirmed her results, during the year of 1953. Finally she showed her results to Noble. He listened to her politely without saying much, and then he sent samples out of the hospital laboratory for confirmation.[112]

In addition to completing her assignment by showing that *vinca* had no effect on blood glucose, Halina proved through her own initiative that it *did* have an effect on white blood cells – and a powerful one at that: it killed them. Her discovery meant that *vinca* had potential for use as a cancer treatment.[113]

Noble's underwhelming response must have come as something of a relief to Halina since she had acted without authorisation. At the same time, she did not take offence at his off-handedness. Conscious of her precarious position, 'she respected his desire for accuracy and she thought that his listening to her at all was a mark of tolerance'.[114] She went on with her work at the lab and at some point suggested that the team perform a technique called chromatography to separate and isolate the active component of *vinca*. The suggestion was ignored. Fortunately, Halina had other things on her mind. On 1 August 1953, she married James Russell Robinson, who also worked at the Collip Laboratory. When she left the lab in 1955 or 1956 for a new position elsewhere, she gave Collip her laboratory notebook, which documented her discovery.[115]

Years later, Halina attended a speech given by Noble, by then famous for the discovery of vinblastine, a breakthrough cancer drug derived from Madagascar periwinkle that is particularly effective at treating leukaemia and lymphoma.[116] When asked why he had decided to analyse the white blood cells, he said he did not know. Halina, sitting in the audience, did know. But, as she told Duffin in 1999, she 'did not have the nerve to speak up'.[117] She eventually worked up the courage to write a journal article, published in 1991, in response and as a correction to an article Noble published in the same journal the previous year about the discovery of vinblastine.[118]

While Noble had thanked Halina and others in the Collip Laboratory who had 'assisted in the research' in a 1958 academic paper about the potential uses of *vinca rosea* in chemotherapy, the title of the paper – 'Role of Chance Observations in Chemotherapy: *Vinca Rosea*' – undermines the fact that there was nothing accidental about Halina's experiments.[119] Her decision to test the rats' white blood cell counts was deliberate. That the experiments were not directed by Noble but initiated by a woman laboratory technologist may account for why Noble and his male colleagues chose to account for it as 'chance' and then later forgot the impetus altogether. Halina herself did not reject this possibility.[120] Some proof of this may lie in Noble's erroneous attribution of Halina's observations of the rats' white blood cells to a male scientist, Dr Vernon Colpitts, who later encouraged Halina to write the corrective article published in 1991.[121]

Her suggestion before leaving the Collip Laboratory that chromatography be used to isolate the active component of *vinca* had also been prescient. The technique was eventually used for this purpose, but only after chemist Dr Charles Beer (1915–2010) joined the Collip Laboratory in 1954.[122] Beer's efforts led to the successful isolation and purification of vinblastine (also called VLB) in 1958, thus enabling its use as a chemotherapy drug. The first clinical trials of vinblastine as a treatment for patients with Hodgkin's disease, a type of lymphoma, were conducted the following year. It continues to be used today for various types of cancer, especially Hodgkin lymphoma and testicular cancer.[123]

In 1997, Noble and Beer were inducted into the Canadian Medical Hall of Fame for their joint discovery of vinblastine. Although Noble had never set out to identify a cancer treatment and probably would not have without Halina's discovery, her important role was not recognised as part of the induction. Noble died in December 1990, and went to his grave without acknowledging (or perhaps even remembering) the importance of Halina's rogue experiments, although Beer had helped Halina prepare her 1991 article about her part in the discovery.[124] It has been thanks in large part to Jacalyn Duffin's interventions that Halina's role is now a part of the story, including on the website of the Canadian Hall of Fame.

A Weak Cocktail

Had these women given up their research in the face of the dismissals they experienced during their careers, they might never have made the discoveries that have contributed so much to improving life and health for humanity and the natural environment. Of course, had they not survived the Second World War and the Holocaust, they would never have had the chance to persevere. But while surviving the Holocaust and the war against many odds can be attributed primarily to luck, their successful careers against many odds can only be attributed to the kind of perseverance and patience that Ruth Pick referred to in the quote that opened this chapter. Despite dismissals and doubts, these six women never gave up in frustration.

When the 1950s served these women its characteristic cocktail of progress and regression, it may have left bitterness, but it could not defeat them. Having met with and survived genocide and ethnic cleansing during the war, they had developed some degree of resistance, if not immunity, to the evils of the gender bias and discrimination. Having seen a glimpse of their futures, it is time to go back to the past to understand the trauma and tragedy that shaped their lives and careers.

Chapter Two

Interrupted Lives

> I should thank Mussolini for having declared me to be of an inferior race. This led me to the joy of working, not any more unfortunately in university institutes, but in a bedroom.
>
> Rita Levi-Montalcini, MD, 2009[1]

When Rita Levi-Montalcini spoke these words, the Second World War and the Holocaust were many decades in the past and she was a Nobel Laureate. One hundred years old at the time, she was nonetheless as mentally sharp as ever. While not the ramblings of an old woman, some might be inclined to think that she was nonetheless viewing these terrible events and her firsthand experiences of them through glasses tinted rose-coloured by great accomplishment and international acclaim. Others might be shocked that she could so blithely attribute something positive to the Fascist racial laws that led to millions of deaths during the war. Unlike many of those victims and millions more who survived, Rita was never imprisoned in a ghetto or concentration, labour, or death camp. One might venture that if she had been, she might not have made such a statement. These potential interpretations aside, Rita's quote reflects her personal experiences of surviving the Second World War and the Holocaust as a Jew in Italy, which she did in part by finding comfort and beauty in her medical research when all around her was uncertainty and horror.

In distinct ways, each of the women in this book were STEM pioneers at least in part because of their experiences during the Second World War and the Holocaust. Their lives were interrupted and torn apart by these cataclysms. What their lives might have been like had the war never happened is anyone's guess. Certainly, they would have been better in many ways, not to mention more peaceful. But the war did happen, and it shaped the women personally and professionally. Among those fortunate enough to survive, they also managed to successfully parlay their experiences into careers that have benefited all of humanity. As Rita's quote suggests, they all found joy in their work, even when it was under terrible and dangerous conditions during the war, whether hiding, in ghettos, or in camps.

Although Rita's words can be interpreted in a variety of ways, her comment about thanking Mussolini can best be read as derision – of Mussolini, the racial laws, the notion that Jews are inferior, and the ultimate failure of Fascism in the 1940s. Even at the age of 100, even as a Nobel Prize winner, she took immense pleasure in the fact that the Italian racial laws were what made her important research during the war possible in a way that it might not have been otherwise. In other words, Mussolini and the racial laws had spectacularly backfired.

This chapter goes back to the early years of the Second World War, starting with the effect of racial laws on Jews studying and working in medicine and science in Germany, Italy, and Czechoslovakia, and how these laws affected Ellen Loeb Katz, Ruth Pick, and Rita Levi-Montalcini. The chapter continues along with the progression of the major events of the period and how these were experienced by and affected all six women personally up to the eve of the Final Solution.

Living Space

In 1933, the year Adolf Hitler and the Nazi Party came to power, around 3,500 Jews lived in the city of Wuppertal in western Germany. Ilse Loeb, as Ellen Loeb was known in her youth, and her family were among them.[2] Wuppertal was and still is an industrial town, home to recognisable companies like pharmaceutical giant Bayer. Ellen's father, Julius Loeb, was a physician who had served Germany in the First World War. Her mother, Dina, was a dietitian. The couple's first child, Gertrude (Trudy), who was born in 1914, later recalled that she saw her father for the first time in 1918, after his honorary discharge from the military. On 30 January 1920, Ilse (Ellen) was born, the final child for Julius and Dina. The family lived in a spacious ten-room apartment and employed a maid. Julius was president of the Wuppertal synagogue, which had been established in 1865. In July 1934, he was awarded with the Iron Cross by the Nazi-controlled government for his military service. But the Loeb's comfortable life was crumbling around them even then. They and every other Jew in Wuppertal and Germany were in danger.[3]

Initially at least, Adolf Hitler and the Nazis had no intention of addressing the so-called Jewish question by murdering German Jews. They just wanted them out of Germany. To facilitate this, a slew of anti-Jewish legislation was enacted in the 1930s that restricted the rights and freedoms of German Jews, including the right to work in professions like law and medicine and attend university. Julius Loeb's status as a war veteran initially exempted him from the laws that prohibited Jews from working as physicians. Nonetheless, he couldn't help but realise the danger the ascent of the Nazis posed to his and other Jewish families. Through his role at the synagogue, he began helping Jewish teenagers leave Germany. One of those he evacuated was 17-year-old Trudy Loeb, who was sent to live in Amsterdam.[4]

Life for the other members of the Loeb family in Germany went from bad to worse. Around the time that Julius was awarded the Iron Cross, his medical licence was revoked under dubious circumstances. Unable to work, his wife discovered him contemplating suicide. So, in 1935, Julius joined Trudy in the Netherlands while Dina and 15-year-old Ellen, who was finishing secondary school, stayed behind to wrap up the family's life in Wuppertal before also emigrating to Amsterdam. Three years later, in early November 1938, when the Loeb family was settled in Amsterdam, the synagogue in Wuppertal burned to the ground during the chaos and anti-Jewish violence of *Kristallnacht*, the Night of Broken Glass, also known as the November Pogrom. Many of the remaining Jews in Wuppertal were eventually deported to Nazi concentration camps. Few survived the Holocaust.[5]

Kristallnacht was not a single pogrom – a word historically used to describe anti-Jewish violence and mass murder – but a wave of organised pogroms committed throughout the German Reich on 9 and 10 November 1938. It was ostensibly in retaliation for the assassination in Paris of a Nazi official by a young Polish-Jewish man, Herschel Grynszpan. By that time, Nazi Germany had begun to build its empire, or Reich, by annexing Austria and an area known as the Sudetenland, and invading Bohemia and Moravia. Behind this territorial expansion was the ideological principle of *Lebensraum* (literally, living space). Like the ideology of 'Manifest

Destiny' that encouraged westward expansion of the United States in the nineteenth century, *Lebensraum* was expansion in the name of nationalism and a supposedly superior race of people. It justified the German takeover, by any means necessary, of countries that were believed to rightfully belong to Germany to perpetuate the supposedly superior Aryan race and cleanse them of 'inferior' non-Aryans, such as Jews and Slavic peoples. As part of this ethnic cleansing, thousands of Polish Jews living in Germany were forced out of the country in late October 1938, including Grynszpan's family. As Jews, however, they were also barred from entering Poland, forcing them into a desperate existence in a refugee camp in the town of Zbaszyn, in the border region between Germany and Poland. Aware of the terrible circumstances facing his parents and other refugees, Grynszpan gained access to the German embassy in Paris, where he shot embassy official Ernst vom Rath, who died two days later. The anti-Jewish violence that followed during *Kristallnacht* resulted in the destruction of Jewish property and murder of Jews not only in Germany, but also in Austria and Czechoslovakia.

In Amsterdam, the Loeb family was spared from the violence of *Kristallnacht*, but the shock of the news about the destruction in Germany nearly killed Julius Loeb, who suffered a massive stroke.[6] Although he survived, life in the Netherlands was becoming increasingly precarious for the Loebs and other Jews. Trudy left Europe with her husband in August 1939 and settled in Dallas, Texas. Ellen, then 19, stayed behind in Amsterdam with her parents. They tried at least once to leave the country but were not successful. In May 1940, the Nazis invaded and successfully conquered the Netherlands, trapping Ellen and her parents.

* * *

Ruth Pick (née Holub) was born on 13 November 1913, in what was then called Carlsbad, a spa city in Bohemia that was part of the Austrian Empire. Her father had served in the Austrian army during the First World War. Following the war, Carlsbad was renamed Karlovy Vary when it and other parts of Bohemia were incorporated into the newly formed country of Czechoslovakia. Despite the change, most residents of Karlovy Vary were ethnic Germans, including – technically speaking – the Holub family. Ruth grew up speaking German at home and school. Although she learned Czech at school, she could barely speak the language when she began attending medical school in Prague in 1932.

Jews had contributed to developing major industry in Carlsbad/Karlovy Vary during the nineteenth century, but the Jewish population was relatively small in the 1930s. The Holub family were secular Jews. They celebrated Christmas, which Ruth's father felt was a pagan feast rather than a Christian one, but not Hannukah. The family lived comfortably in a large apartment provided by the porcelain company in Karlovy Vary where Ruth's father was head of foreign sales. During her youth, Ruth enjoyed studying physics, and eventually enrolled in medical school at the German University of Prague with her father's encouragement. In 1935, Ruth met Alfred Pick, already a physician, while doing her internship at one of the university's two German medical clinics – the one that employed Jews.[7]

Ruth's father died not long before she graduated from medical school in April 1938. One month later, 24-year-old Ruth married Alfred Pick in a low-key ceremony. Her father's death and her mother's related depression subdued these otherwise joyous occasions. The Picks, like many others in Czechoslovakia, were also alarmed by the Nazi annexation of Austria in March 1938, known as the *Anschluss*. From the German point of view, the German-speaking regions of Bohemia (including parts of Czechoslovakia), Moravia, and Silesia – known collectively as the Sudetenland – that had once been part of the Austrian Empire belonged to Germany. Like Austria, these lands and their German-speakers (at least the non-Jewish ones) were *volk* – ethnic Germans who needed to be brought back into the German Reich. With Austria subsumed under the German Reich, Hitler set his sights on the Sudetenland as the next phase of the *Lebensraum* expansion. Aware of the possibility of a German invasion of Czechoslovakia, Czech troops were mobilised in May 1938. 'We all got very tense,' Ruth later explained. 'Because it came very close. And if they can take Austria, they can come into Czechoslovakia.'[8]

With Hitler threatening war unless the Sudetenland was ceded to Germany, European leaders signed the Munich Agreement on September 30, 1938, carving up Czechoslovakia and effectively sacrificing not just the Sudetenland but, ultimately, all of Czechoslovakia in the process. As historian Mark Mazower has written, 'it was a disaster for the Czechs and a catastrophe for all those hoping to stem the German drive to war'.[9] But that was in the future. For the moment, Britain's prime minister Neville Chamberlain proclaimed, 'peace for our time'. Having been born and raised in the Sudetenland, Ruth knew better. When the Munich Agreement was signed, she and Alfred were working at a sanatorium in an area of Silesia that was part of the Sudetenland. As Jews, they were thus in immediate danger. Following the annexation of Austria, thousands of Austrian Jews had been sent to Dachau concentration camp in Germany. Conscious of the possibility that the same would (and did) happen in the Sudetenland when the Nazis annexed it in October 1938, the Picks fled to Prague, where they moved in with Alfred's parents.

Even in Prague, many were aware that it was only a matter of time before the Nazis took over the rest of Czechoslovakia. 'We knew what was going to come,' Ruth explained late in her life. 'We didn't know how badly it was going to go.' The Picks considered emigrating to England but decided against it due to their lack of funds and knowledge of the English language. As time went on, more of the Picks' family and friends emigrated, but the Picks couldn't secure a visa. By March 1939, when the Nazis occupied Prague, it was too late.[10] Anti-Jewish regulations were soon imposed on the Czech Jews, who were bound by curfews, forced to live in crowded accommodations, and required to always wear the yellow Star of David, identifying them as Jews.[11] 'It was a very tense time. I can't even make you feel it,' Ruth explained. What she realised only in retrospect was that their situation was only going to get worse.[12] The Nazi conquest of Austria, the Sudetenland, and Czechoslovakia was just the beginning of the Nazi quest for *Lebensraum*. More living space had to be found for the German *volk*, and in that space, more room would have to be made by cleansing the German Reich of Jews, Slavs, and others.

The Destruction of Poland

In her memoir, Halina Czajkowska described her Polish childhood as heaven. An only child, born on 23 June 1927, she was adored by her parents, Czesław and Emilia. With her father in the Polish military, the family moved frequently. By the time Halina was five, they had moved from her birthplace in Jarosław, southeastern Poland, to Kielce, in south-central Poland, and, finally, to Berezno, in eastern Poland bordering Ukraine.[13] Even many years later, Halina recalled with delight taking a horse-drawn carriage from a train station to her new home in Berezno: 'The weather was warm and sunny that day and the flat countryside was absolutely beautiful, and your eyes could roam for miles over the sandy landscape. From time to time, we passed through forests where we enjoyed the fragrant pine-scented air.'[14]

Berezno shaped Halina for the next five years of her life. She was happy at school, especially in her drawing classes, was fascinated by the work of the farmers, went sledding in the winter, and enjoyed religious holidays. 'Berezno means so many things to me,' she wrote in her memoirs.

> I remember the sweet smell of burning incense during May's evening services to Holy Mary in our lovely white church. I remember the autumn evenings, with the melodious songs of the women in the fields at the end of the day as they gathered the potatoes, which exuded such a delicious smell and tasted better than anything else in this world.[15]

Berezno was Halina's 'paradise'. In a way, it was also a paradise for Jews. Before the war, more than 90 per cent of the population of Berezno was Jewish. One Jewish resident of the interwar period likened Jewish life in Berezno to 'a peaceful river'.[16] It is no wonder then that Halina, who was Catholic, had meaningful encounters with some of the many Jews who lived in the town. In her memoirs, she recalls experiencing with wonder a kosher kitchen for the first time, and her curiosity about this arrangement. And she writes longingly of favourite foods, many of which were purchased by her mother from local Jewish businesses.[17]

The entire family regretted leaving Berezno in 1937 to move to Zamość, where Halina's father took up a new position as Lieutenant Colonel in the Polish army. Like Berezno, Zamość had a large Jewish population – around 45 per cent of the total inhabitants. As a non-Jewish Pole and the daughter of a Polish military officer, Halina lived a privileged and somewhat sheltered life. She would not have been fully aware of the antisemitism Jews faced in Poland or of just how rare the tranquillity of Jewish life in Berezno was.

But her path in life put her in many of the places that Jews lived in large numbers and that were consequently targeted by the Nazis with intense ferocity. Sheltered as she was, she was almost certainly unaware of how dangerously close the end of life was as she and all Poles – Jewish and non-Jewish – had known it. In particular, she had no way of knowing that the thriving Jewish communities she had encountered in Berezno and Zamość were about to be annihilated.

Hitler may have had promised peace, but he had no intention of stopping with the annexation of Austria and the Sudetenland and the occupation of Czechoslovakia. The latter was especially strategic as it was his gateway into Poland. Whereas Hitler had intended to make Austria and Czechoslovakia part of the German Reich with as little disruption as possible, his goal in Poland was nothing short of destruction.[18] The Poles were to be subdued, enslaved, and, if necessary, murdered, to make room for the flourishing of the German empire and race. Non-Jewish Poles who identified as Germans – *Volksdeutsche* – might be saved. But no such concessions were reserved for Jews.

In cooperation with Soviet leader Joseph Stalin, Soviet and Nazi forces invaded Poland on 1 September 1939. Within days of the joint invasion, Halina's family home in Zamość suffered extensive bomb damage during a Nazi air raid, while Halina, her mother, and a cook were inside. 'First, we heard an explosion, and seconds later, a strong blast hurled us against the wall, followed by another blast that threw us in the opposite direction and down onto the floor,' described Halina, who was 12 at the time. 'Plaster fell from the ceiling and the air became so white with dust that we could barely see.'[19] Incredibly, no one in that home was seriously injured, though hundreds of others were hurt in these and other bombing raids on Zamość.[20]

While Halina's father went off to fight the invasion, she and her mother were caught up in alternating occupations of Nazi and Soviet forces. Zamość was occupied first by Nazi troops, then by Soviet troops, and then again by the Nazis – all in a matter of weeks. The Soviet occupation of Zamość enabled thousands of Jews to flee. But the return of the Nazis was deadly for both Jews and non-Jews. Polish intelligentsia – teachers, religious leaders, political activists, judges, etc. – were murdered by the tens of thousands throughout Poland, including in Zamość. As for the Jews, Halina began to observe the literal and figurative signs pointing to their fate all around her in the winter of 1939/1940. 'The first glimmer of trouble had come in the puzzling order for Jewish people to wear armbands, but so many other horrible and confusing things were to happen that winter,' she wrote. In early 1940, 'we heard about mass arrests and saw placards calling on the Jewish people to report, with their belongings, to a certain location in the city. They were told that they would be transported to a part of Poland where they would have autonomy.' She described how, at the time, many non-Jewish Poles felt envious of the Jews who, they believed, were being 'removed from the terror of that city'. In retrospect, she lamented at 'How little we knew, at that time, about what the Germans were really planning for the Jewish families!'[21]

One of these Jewish families was Halina Neujahr's.

* * *

Born in Warsaw, the capital of Poland, on 28 October 1924, Halina Neujahr was the third and last child, all daughters, of David and Gitla-Ejdla Neujahr. The Jewish population of Warsaw comprised around one-third of the city's 1.3 million inhabitants, making it the largest Jewish population in both Poland and Europe, and second in the world after New York City.[22] The family lived comfortably in the city's Jewish quarter, but David Neujahr, who worked in the textile business, saw only a gloomy future for Jews in Poland.[23] While Halina Czajkowska's life in Poland before the Second World War was paradise, the same cannot always be said for Jewish families like Halina Neujahr's. The antisemitism that had been a reality for Jews in Warsaw and elsewhere in Poland was increasing in the 1920s and 1930s. 'The Jews were an urban element that the Poles had long wanted to undermine and deprive of a livelihood – all the more so in light of the persistent economic difficulties and severe depression that overtook Poland between the wars,' according to Polish historian Yisrael Gutman.[24] Of course, not all non-Jewish Poles felt this way. Rather, this reflected – at least in Halina's understanding – her father's experience of life in Warsaw.

Envisioning a better life for his family, David Neujahr travelled to New York to lay the groundwork for the rest of the family's emigration from Poland. In February 1929, just months before Halina, her sisters, and their mother were to sail to New York to join him in New York, David was killed in an elevator accident. Halina was 4 years old.[25] She and the other Neujahr women stayed in Warsaw, where they lived comfortably thanks to her late father's business, an insurance settlement related to his death, and good financial management by Gitla.[26]

Ten years later, on 23 August 1939, 14-year-old Halina was enjoying summer vacation at her aunt and uncle's home in Gostynin, northwest of Warsaw, when the Polish army was mobilised to defend against the anticipated Nazi and Soviet invasions. Halina later recalled,

> We did a lot of swimming, and I remember that on the day I made my greatest swimming achievement – I swam across a dangerous lake, which was rather deep, and it was about three kilometers [across] altogether – came the mobilization order and everything changed, all those joyful occupations we young people had [ended].[27]

Halina's eldest sister, Rachela, was living in England, while her other sister, Sara, and mother Gitla were in Warsaw. Rather than return to Warsaw for the start of school on 1 September, Gitla advised Halina to stay where she was, assuring her that things would soon settle down. Instead, the Second World War began after the Nazi invasion of Poland on 1 September prompted the United Kingdom and France to declare war on Germany. In Gostynin, Halina watched as Polish soldiers injured by the Nazi forces during the Battle of the Bzura near the town of Kutno made their way back to Warsaw. The Nazis followed in their wake, occupying Gostynin on 16 September. Soon after, Halina recalled, all non-military men in the town age 15 and older, including Halina's uncle and 15-year-old cousin, were locked inside a church for three days and nights without food, water, or sanitation. In the end, they were released. But it was only a temporary reprieve. Halina managed to leave Gostynin in October and return to her mother and sister in Warsaw.

While she was away, Warsaw had been under siege for nearly a month before the Polish army surrendered on 28 September and the city was subsequently occupied by the *Wehrmacht* – the unified armed forces of Nazi Germany – on 1 October. After Hitler and Joseph Stalin divided Poland between them, Nazi Germany directly annexed most of the country except for an area called the General Government that included Warsaw and was intended as an administrative zone run by Nazi Hans Frank as Governor-General. Hitler's plans for his newly acquired territories included the '"physical annihilation" of the Polish population'. This included both Jews and non-Jews.[28] But, although murder was certainly one way of accomplishing this, the Nazis realised it was not exactly the most practical way of dealing with eleven to twelve million people. Instead, talk about annihilating both Jews and Poles had co-existed with plans for small numbers of Poles and Jews to live in German-controlled ethnic borderlands.[29]

Once Poland was in Nazi hands, a clearer picture began to emerge of the fate of non-Jewish Poles, who 'were to be turned into a minimally educated slave-labour force for the Germans'.[30] Less immediately clear, at least in late 1939, was how to address the 'Jewish question'. But this would soon become clearer as the occupation of Poland, and the start of the war sparked a change in Nazi mentality that 'provided the Nazis with an opportunity to manifest their racist ideology and implement programs that were the antithesis of accepted mores and deep-rooted humane and cultural traditions'.[31] The desire for the total domination of Poland and its inhabitants, including millions of Jews, became an integral part of the path to genocide.

Nazi Allies

When the Second World War began, following the joint invasion of Poland in September 1939, the Soviet Union was not the only ally of Nazi Germany. In May 1939, Fascist Italy had signed the Pact of Steel with Nazi Germany, in which each country had committed to aiding the other in the event of war. A year after the invasion of Poland, in late September 1940, Italy and Germany reaffirmed their alliance, together with Japan in the Tripartite Pact. Later that year, Hungary and several other countries joined the alliance, followed by other nations in early 1941. Together, these countries formed the Axis powers that fought against the Allied forces, including the United Kingdom, France, and Canada initially, and eventually also the United States and Soviet Union.

When a country joined the Axis alliance, it did not necessarily involve sudden and radical change for its citizens. Rather, in most cases, countries joined the Axis powers because their governments were already aligned with the Nazis' ideologies and ambitions. One area of consensus, at least among the Axis powers in Europe, was that Jews were an inferior race and should be excluded from general society. Before joining the Axis, Italy and Hungary had already established laws restricting the civil rights of Jewish citizens. Miklós Horthy, Regent of the Kingdom of Hungary, was a self-professed antisemite who needed little encouragement from Hitler on such matters. While allegedly not as antisemitic as Horthy, Benito Mussolini, Italian Prime Minister since 1922 and dictator since 1925, supported the promulgation of Aryan racial theories and instituted racial laws at least partly to impress Hitler.[32] Where Mussolini and Horthy differed from Hitler was in exactly how to handle the so-called Jewish question. These differences of opinion, together with Hungary and Italy insisting on maintaining self-determination regarding to how to deal with the Jews in their own countries, played an important role in the lives of Rita Levi-Montalcini and Eva Klein.

* * *

Dr Rita Levi-Montalcini (or Dr Rita Levi as she was then known) was attending a scientific conference in Sweden when news of the Nazi invasion of Poland reached her.[33] Instead of returning to her native Italy, she took the last civilian ship to Brussels, Belgium, where she had been living and working since March 1939. Rita left Italy because of the increase in antisemitic restrictions on the country's small Jewish population during the 1930s, culminating in the racial laws of 1938. Among other restrictions, the laws prohibited marriage between Jews and non-Jews, and Jews were excluded from working in public office, at universities, and in all professions. Born into an affluent Sephardic Jewish family in Turin on 22 April 1909, Rita was consequently dismissed from her academic and research positions at the University of Turin in northern Italy and was banned from practising medicine.

Still early in her career (she had completed her medical degree just three years earlier, in 1936), 30-year-old Rita was determined to at least continue her research. In direct defiance of the racial laws, she continued her research into the embryonic development of the nervous system that she had been pursuing in collaboration with non-Jewish neurophysiologist Dr Fabio Visintini at the University of Turin. As a student, Rita had become skilled in histological methods of staining nerve cells for microscopic study under the instruction of Professor Giuseppe Levi. One of these methods was silver staining, or Golgi staining, a technique developed in the late nineteenth century which, as Rita explained, 'makes nerve cells stand out dramatically in the smallest detail, dark against the golden-yellow background of surrounding tissues'.[34] Not only had she adopted this method, she had also modified and mastered it early in her career.[35]

Rita found the work exciting and rewarding, but the danger involved was increasing. 'We carried out our research in an atmosphere every day more menacing as a result of the virulent newspaper articles saluting the racial campaign for opening the eyes of Italians,' she later recalled.

> By March 1939, I could no longer attend university institutes without risking denunciation or endangering Aryan friends, who would have become guilty of works of pietism – the accusation brought by the regime against those who manifested sympathy or solidarity with Jews.[36]

So, she took up an offer to work at a neurological institute in Brussels, where her eldest sister and her family were also living. But the war loomed over all of Europe, including Belgium, where the military was mobilising in preparation for a Nazi invasion. In December 1939, Rita, accompanied by her sister and her family, returned to Turin. Despite the racial laws, Jews in Italy did not face mortal danger like they did under Nazi rule. Back in Italy, she initially practised medicine, treating former patients from the University of Turin Medical Clinic, where she had worked before the racial laws were enacted. 'These poor people, who lived in the attics of houses in old Turin, did not care about the laws and were glad of my visits and the help I could offer them within the limits of my scarce finances,' Rita wrote in her memoirs.[37] But this work put her in danger of discovery due to her dependence on non-Jewish physicians for their signatures on prescriptions for her patients. The situation forced her to stop practising medicine.

While Rita was left in limbo, Italy formally declared war on France and the United Kingdom on 10 June 1940. Soon after, Italian troops invaded France in support of the Nazi invasion, which had begun in May. Before the end of June, France had fallen.

* * *

Whereas Italy's approximately 48,000 Jews comprised only a fraction of 1 per cent of the country's prewar population, Hungary's Jewish population before the war was close to 450,000, around 5 per cent of the total population.[38] In Central Europe before the war, Hungary had the second-largest Jewish population after Germany. Around half of the country's Jewish population lived in Budapest, where they made up around 23 per cent of the population.[39] Following the German annexation of the Sudetenland and invasion of Czechoslovakia, Hitler gave Hungary part of Czechoslovakia that had once belonged to Hungary. Another gift of conquered land, part of Transylvania, followed in August 1940. Already informally allied with Nazi Germany and Fascist Italy, Hungary formally joined the Tripartite Pact in November 1940. The following spring, Hungary and Germany invaded Yugoslavia and then divided it between them. In this way, Hungary gained not only huge swathes of land before and during the Second World War but also hundreds of thousands more Jews, bringing the country's Jewish population to around 800,000.[40]

Eva Fischer – as Eva Klein was known before her marriage – was a 15-year-old student when Hungary formally joined the Axis alliance in November 1940. She had dreams of attending university, though she knew that her chances of doing so were slim. In 1920, soon after coming into power, Hungary's Fascist and antisemitic head of state, Miklós Horthy, established a *Numerus Clausus* law to restrict the number of Jewish students attending Hungarian universities to no more than 6 per cent of all students.[41] In other respects, Eva's early life was happy. She was the third and final child born to middle-class Jewish parents on 22 January 1925, in Budapest, Hungary's capital city. Known during her youth as Baba (which means baby), Eva was cared for by a devoted non-Jewish Polish nanny. She attended school with both Jews and non-Jews and even had a non-Jewish boyfriend. In addition to an active social life, Eva was a good student and participated in sports and extracurricular activities like drawing and theatre.[42]

Life for all Hungarian Jews became increasingly difficult starting in the spring of 1938, when the Horthy government began enacting a steady stream of anti-Jewish legislation. 'Between May 1938 and March 1944 the Hungarian Parliament passed 22 antisemitic laws,' according to the United States Holocaust Memorial Museum. 'An additional 267 anti-Jewish ministerial and governmental decrees were issued during the same period.'[43] As had happened in Italy, Hungarian Jews were restricted from working in most professions and marriage between Jews and non-Jews was banned. In addition, the Hungarian Labor Service was established in 1939 to draft Jewish men between ages 20 and 48 into forced labour.[44]

Even as a teenage student, Eva was affected by the antisemitic legislation. Smart and ambitious, she nonetheless struggled with an internalised sense of inferiority that was the direct result of the social and political treatment of Jews and the antisemitic racial laws. With a wide variety of interests but no expectation of attending university, she struggled to identify her calling in life.[45] Her social life also suffered. Friendships with non-Jewish students became difficult to maintain when Jewish students were banned from participating in school sporting competitions, driving a wedge between the Jewish and non-Jewish students. Compellingly, however, the political instability in Hungary actually helped to bring Eva and her group of Jewish and non-Jewish friends closer together. Some of these friendships were maintained long after the war. One contributed to saving her life during it.[46]

* * *

As the Second World War escalated in the summer of 1940, Eva Fischer and Rita Levi-Montalcini were undoubtedly living in difficult circumstances. But whether they knew it or not, their circumstances in Hungary and Italy were better than those in Nazi-occupied Europe, including Ellen Loeb in the Netherlands, Ruth Pick in Czechoslovakia, and Halina Czajkowska and Halina Neujahr in Poland. In fact, in the early years of the war, the Jewish populations of both Hungary and Italy increased as Jews fled the Nazis in other parts of Europe, including the Netherlands. With most of unoccupied Europe, the United States, and other countries severely restricting Jewish immigration, Hungary and Italy seemed a lesser evil than Nazi-occupied Europe. Nonetheless, any sense of security that Jews in Hungary and Italy might have felt was sorely misplaced. Mussolini and, to a greater extent, Horthy were less concerned with protecting Jews from Hitler than with protecting their autonomy within their borders and reserving their right to decide the fate of the Jews and others within them.[47]

Education as Resistance

In Hungary, Eva Klein was able to continue her education. But the Nazi and Soviet invasion of Poland had put an end to formal education for most young people, including Halina Czajkowska and Halina Neujahr. Before the invasion, 12-year-old Halina Czajkowska attended a school at the Zamość military base alongside Jewish, Catholic, and Protestant students. She was involved in a variety of extracurricular activities, including Scouts and the Red Cross, and took private music lessons.[48] Halina Neujahr, age 14, had been attending a private school in Warsaw attended only by Jewish students but with both Jewish and non-Jewish teachers.[49] By their own accounts, both Halina N. and Halina C. liked school and were each planning on starting a new school year in September 1939. Those plans were shattered with the destruction and chaos of war. By late fall 1939, Zamość and Warsaw – home to Halina C. and Halina N., respectively – were under the jurisdiction of Nazi Germany. Thousands of Polish intellectuals, including teachers, had been executed or arrested and sent to concentration camps. Polish secondary schools and universities were closed, and teaching in Polish was banned, as were the subjects of history and geography. Poles were to become nothing more than uneducated labourers for the Nazi Reich.[50] Under these conditions, education became more important than ever for Poles, who arranged clandestine education despite the great danger of doing so.

In early 1940, Halina Czajkowska's mother enrolled her in the highest grade of a private junior school in Zamość. The danger involved in this meant that special care had to be taken to avoid detection. 'We were advised to walk to and from school alone and at staggered times in order not to attract attention,' Halina wrote. At the same time, private teaching groups were organised for older students in private homes. These were even more dangerous, since upper secondary education was completely banned by the Nazis. 'That plan did not last long for the Gestapo broke in on some gatherings and arrested everyone: children, parents, and teachers,' she explained. 'After interrogation at the Gestapo Headquarters about who had organized the schools, all the prisoners were sent as slave labourers to Germany.' Within a few months, such efforts ceased, including the school Halina had been attending. In parting, Halina's teachers gave her exercises to do at home, which she took very seriously.[51]

On her own, Halina C. soon graduated to upper secondary school textbooks, through which she was learning Latin and other advanced subjects. Late in 1940, she and her mother left Zamość and settled in Warsaw. There, it was agreed that a new attempt would be made to enrol Halina in an underground school. Since trade schools were allowed by the German occupiers, the school Halina attended was officially a two-year girls' secretarial school. In reality, it was a Polish high school for girls operating with the knowledge and support of the Polish underground resistance group the Home Army (AK). Being a part of this operation in any way was extremely dangerous, and the pretence that it was nothing more than what it claimed to be was critically important. As Halina later recalled:

> We had to observe strict rules, especially to not carry any books in the streets; notes had to be folded and put out of sight in our pockets. We had a lot of homework, which had to be reported verbally from memory. Any slip from this constant vigilance was risky, not only for teachers and students but also for our entire families.[52]

The consequences of such a school being discovered were devastating. One of Halina's contemporaries, non-Jewish Pole Anna Krasińska, was just 10 years old when her school was raided by the Nazis in April 1940. In 1946, she described what happened:

> We were at school in Gdańsk. German secret policemen came and took the teacher because she was in AK [Armia Krajowa (the Home Army), the primary Polish resistance group]. They shot her dead right then and there. The date was 19 April 1940. They took us to some motor vehicles and drove us to the railway station. The drive took seven hours. There were a lot of elderly women with us in the cattle wagons (each wagon held a hundred people); our group was made up of fifty young people, including nineteen boys. Many people starved and suffocated along the way – thirty-three in our group made it to Neustrelitz. The boys were separated from us, and we girls were taken to a single

cell in a prison. For two days, we ate nothing. The prison was overcrowded. In the prison courtyard, women were cooking food in pots over fires. When one of them stepped away, we took some food. Inside the pot, we found human ears and fingers. We ate it because we were extremely hungry. The meat came from people who had died at the prison. Two days later, we were fed swede [rutabaga] soup and escorted to a factory. Before leaving, we had been made to change into striped prison uniforms. At the prison, they beat us badly.[53]

* * *

Anna's experience is a vivid example of how the Nazis stripped not only education from young Poles, but also childhood. By her own account, Halina Neujahr's childhood ended in late August 1939. After returning to her mother and sister in Warsaw in October 1939, Halina's daily life became centred on meeting the family's basic needs. The Nazis systematically confiscated Jewish property, businesses, and bank accounts. Radios were confiscated, as were furs. Loath to let the Nazis have these items, the Neujahrs smashed their radio and stuffed the furs up the chimney. Halina's mother had also managed to hide other valuables, which she sold on the black market in exchange for food and other necessities. Their apartment in the Jewish area of Warsaw became overcrowded as the Nazis began to force Jewish families from other parts of the capital to live in a limited area in preparation for a Jewish ghetto. In addition, Jewish refugees began flooding into the capital from other parts of Poland, and they too were crowded into the developing ghetto. At one time, Halina, her mother and sister were living with as many as eight other people in a twenty-five square metre apartment. A curfew was imposed on Warsaw's Jews and Halina's mother implored her not to travel far from home due to attacks on Jews and random forced labour roundups. Amidst all of this, Halina longed to attend school.[54]

The Warsaw Ghetto was sealed on 16 November 1940, trapping 30 per cent of the population of Warsaw into a space equivalent to just 2.4 per cent of the area of Warsaw. Of the approximately 380,740 inhabitants of the ghetto, most were Jewish.[55] The official reason the Nazis gave for the establishment of the ghetto was to protect the Nazi troops in Warsaw and the general population from disease emanating from the overcrowded Jewish quarter.[56]

Of course, this decision led to mass death within the ghetto, an outcome expected by the Nazis and which contributed to solving the Jewish question through 'gradual extermination'.[57] Nonetheless, because the Nazis were fearful of the rampant spread of disease for their own sakes, they allowed the establishment of hygiene courses for the inhabitants of the ghetto.

With the German occupiers' permission, vocational training courses were established in the ghetto that were administered by the Jewish self-government, the *Judenrat*. Whereas the Nazis viewed these courses as a way of creating specific types of labourers that would be useful to the Reich, the *Judenrat* saw them as vehicles for clandestine education.[58] The hygiene course in the Warsaw Ghetto was in fact a first-year medical school course led by the eminent Polish microbiologist Professor Ludwik Hirszfeld (1884–1954).[59] Halina Neujahr jumped at this chance to continue her interrupted education. In addition to learning practical measures to prevent the spread of epidemics like typhus and typhoid fever, she and the other students were taught bacteriology, chemistry, and histology. As with other clandestine education initiatives, the students had limited access to books and were not allowed to take notes that might give away what the students were actually being taught if discovered by the Germans.

Because the classes were held near to a Nazi guard post along the perimeter of the ghetto, gunshots were often heard during the lectures, as inmates tried to escape. Professor Hirszfeld later wrote, 'From below, I could hear shots and cries of victims. But the students sat looking at me and were completely absorbed in what I was saying.'[60] This was not an exaggeration as far as Halina was concerned. Years later, she recalled the 'wonderful teachers', including Professor Hirszfeld but especially bacteriologist Dr Róża Amzel (1904–1943), who did not survive the Holocaust. 'She was a marvelous teacher, really,' Halina said. 'She spoke with such feeling for those microbes and how they work and how to fight them. And really one could see that these are like human beings.'[61] Without a doubt, the seed of Halina's future career as a biochemist was planted in the terrible conditions of the Warsaw Ghetto.[62] In this description of Dr Amzel, it is impossible not to see Halina herself, especially when she described the beautiful red crystals of B12 that were such an important part of her early scientific research.

* * *

Halina Neujahr and Halina Czajkowska were passionate and dedicated students before the Second World War. If anything, they only became more so after it began. The forced interruption in their formal education caused by the war and the Nazis' determination to turn Poles into a slave labour force for the Reich reinforced to them just how important education was. Young as they were, they understood the consequences if they were caught getting an illegal education. At the same time, they were acutely aware that their lives were already of little importance to the Nazis. So, they resisted and fought, not with guns but with their minds. In doing so, they set themselves on the paths that would eventually lead to them making important contributions to humanity.

Working to Live

Under the racial laws of Nazi Germany and Fascist Italy, Jews were prohibited from working in most professions. The way these two totalitarian regimes enforced this prohibition was markedly different. In Italy, where the racial policies were inspired by the Nazi policies but independent of them, Jews were prohibited from working in professions, including at universities and in medicine. This was why Rita Levi-Montalcini had been forced to abandon both her university research and her medical practice in Italy in 1940. But in some countries occupied by Nazi Germany, Jews were allowed to continue working in certain professions if they provided services only to other Jews. In Czechoslovakia, one of these professions was medicine. This meant that Ruth Pick was theoretically able to continue working in Prague around the same time that Rita was prohibited from doing so in Turin. The vicissitudes of life during the Second World War did not end there. Rather than working as a physician, she was only able to secure legal work as a nurse in a Jewish hospital. Unable to legally work in any capacity in Italy, Rita, on the other hand, ultimately decided to dedicate her energies to scientific research. Regardless of the differences in their circumstances, both Ruth and Rita were working to live. Ruth's work made her useful to the Nazis and increased her chances of survival. Rita's work gave her life purpose and meaning.

* * *

By March 1939, all of Czechoslovakia was under the control of Nazi Germany. The country ceased to exist and instead became part of a new state called the Protectorate of Bohemia and Moravia. The occupiers' intention was to Germanise the land and its people, whom they considered as racially inferior as Slavs. Jews were another matter.

Ruth and her husband Alfred Pick were among the approximately 120,000 Jews living in the Protectorate.[63] Having missed their window of opportunity to emigrate elsewhere, they were trapped in Prague. With the Nazi occupation, 'all the German rules came to us', Ruth later recalled. 'We had a curfew. We had to wear stars. We had to move in with each other. Everybody was allowed only so many square feet in an apartment.' Already living with Alfred's parents, the Picks were forced to take in another Jewish boarder in their small apartment. Jewish movement in public spaces was limited to certain areas and times of day. Food rations were also restricted, far more so for Jews than others. 'We were only allowed to shop from two to four when most of the stores were closed and everything was already sold,' Ruth explained.[64]

Antisemitism had existed in Czechoslovakia before the Nazis. The creation of the First Czechoslovak Republic in the wake of the First World War was accompanied by a wave of anti-Jewish violence, including several pogroms, from 1918 to 1920. Relative peace followed until the Munich Agreement in September 1938 reawakened antisemitism in the country's politics and society. Nazi-inspired racial policies thus began to infiltrate Czechoslovakia even before the Nazis established the Protectorate. Jewish professionals – doctors, lawyers, engineers, etc. – were among the groups targeted before and after the Nazi occupation. Professional organisations like the Czech Medical Chamber began to exclude Jews from membership in the fall of 1938.[65] Under the Nazis, further restrictions were imposed on where and to what extent Jewish professionals could work. In July 1939, doctors and other professionals were prohibited from practising their professions.[66]

Despite the restrictions, Jews in Prague still needed medical treatment. With the city's general hospitals off-limits to them, the authorities allowed the Jewish community to convert an orphanage in Prague into a Jewish hospital.[67] Alfred worked there as a lab technician and Ruth was eventually hired as a nurse. Working twelve-hour shifts, Ruth recalled that she did double duty as both a nurse and a physician. According to her, the male physicians at the hospital were aware that she was a qualified physician and so played cards while she did their work and hers. 'In one hand I had a bed pan and in the other a syringe,' she said with a good-humoured laugh.[68] Their demotions aside, Ruth and Alfred were lucky to have work, not least for the protection it offered them from forced labour, imprisonment, and deportation. In October 1941, just one month after all Jews were required to wear the yellow Star of David, mass deportations of Jews in Prague began. Initially, no one knew exactly where these transports of thousands were headed, though they speculated it was to labour camps.[69]

As the year 1941 ended, more than 10,000 Jews had been deported from the Protectorate, many from Prague. Most had been 'resettled' in places like the Łódź Ghetto in Poland, but some had been sent to the newly established Theresienstadt Ghetto, located in the Protectorate.[70] Those not deported or sent to forced labour suffered under the weight of unemployment, hunger, crowded living conditions, and physical and psychological violence. Disease and illness were rife. At the same time, Jewish health personnel in Prague were few. This made work at the Jewish hospital, where only a relative handful of medical personnel worked, more difficult.[71] But, while Ruth and Alfred were protected by their jobs, at least temporarily, they watched helplessly as Ruth's mother and then Alfred's parents were deported.[72] In addition to the pain of seeing loved ones sent to an uncertain fate, Ruth recalled the psychological turmoil that the transports caused for those left behind. 'You didn't know when your time would come,' she explained. On the other hand, daily life in Prague was terrible, and when Ruth first learned of the Theresienstadt Ghetto she thought it might just be an improvement – a haven where Jews could live free of Nazi violence.[73] It would not take long for her to realise how wrong she was.

* * *

With Italy at war, Rita Levi-Montalcini spent most of her time reading and socialising to alleviate her frustration over being unable to conduct her medical research and practice. In the fall of 1940, a colleague suggested she set up a laboratory of her own where she could continue her research on the nervous system. This 'touched a chord that had been vibrating in me since earliest childhood: the desire to undertake a voyage of adventure to unknown lands', she wrote in her memoirs.[74] Likening herself to the literary castaway Robinson Crusoe, Rita viewed the nervous system as the tropical desert island that she would explore in a makeshift laboratory. The improvised and dangerous nature of this initiative only added to its lustre in Rita's mind. Moreover, she was thrilled that she would be embarking on this adventure in defiance of the restrictions. 'The pleasure I was already savoring in anticipation was enhanced by the prospect of carrying out the project under the conditions contingent on the prohibitive racial laws,' she explained.[75] If the Italian racial laws were the storm that had wrecked her ship and caused her to be stranded on a desert island, Rita was determined to turn a tragedy into an advantage.

Short on money but with support and help from her family, Rita began gathering everything she needed to create her laboratory in her childhood bedroom in Turin. She managed to secure the most expensive items – a stereo microscope and a binocular microscope – and then improvised much of the rest of the equipment, such as an incubator and micro scalpels that she created by grinding down sewing needles. In her earlier research, she had used chick embryos to study the nervous system and so decided to continue to do so since fertilised chicken eggs were fairly easy to obtain. By her own admission, the lab Rita created was crude and reminiscent of the nineteenth century. But, once again, this was part of the appeal as far as she was concerned. 'The "tour de force" of fitting so much apparatus into the small space available to me added to the pleasure of managing to work under prohibitive conditions,' she wrote.[76]

As Rita was conducting her research on chick embryos in the spring of 1941, the tone of antisemitic rhetoric in Italy began to become more vicious. In October, her brother Gino was named in a manifesto along with other 'Jews' (some of those listed were not Jewish) along with the supposed crimes they committed. The manifesto urged that these individuals should be put against a wall and executed with a flame thrower. More generally, Italian citizens were urged to 'fire upon Jews at the slightest suspicion'.[77] Fortunately, this did not happen, and the persecution of Italian Jews was mostly limited to insults and threats under Mussolini's regime.[78] For the moment at least, Rita was thriving and happy in her secret laboratory.

* * *

The racial laws had a profound effect on Jews living in the former Czechoslovakia and Italy. The lives Ruth and Rita had before the war were interrupted, jeopardised, and irrevocably changed. But, although their professional careers were stolen, they both managed to find other ways to continue their work. Working provided them not only with an occupation but also, a tenuous hold onto life. While Ruth's work in the Jewish hospital in Prague effectively preserved her life, Rita's clandestine research in Italy gave her life a purpose. In this way, they were better off than others in their own countries and elsewhere in Europe for whom the destruction of their livelihood left them in a state of despair and desperation. Moreover, because neither Italy nor the Protectorate of Bohemia and Moravia had instituted the same forced labour programmes as Poland, Germany, and Hungary had done, they could live and work without fear of being rounded up on the streets. But 1941 had surprises in store that changed the course of the Second World War and contributed to Nazi Germany's Final Solution to the Jewish Question. These events would soon catch up with Ruth and eventually Rita.

Existing Amidst Chaos

Halina Neujahr, Halina Czajkowska, Ruth Pick, and Rita Levi-Montalcini had each found a sense of purpose during the early years of the Second World War, either through education or work. The same cannot necessarily be said of Eva Fischer (Klein) in Hungary and Ellen Loeb in the Netherlands. Both women were students when the war interrupted their lives, and this rupture destabilised them in ways that they elaborated on surprisingly little in later accounts of their Holocaust experiences.

The main source about Eva's experiences during the Second World War and the Holocaust is the interview recorded by the USC Shoah Foundation in 1997. It contains little information about the early years of the war with the notable exception that the antisemitic *Numerus Clausus* laws in Hungary that restricted the number of Jewish students who could attend university profoundly affected her self-confidence and left her confused about what to do with her life. Despite her uncertainty about her future career and university being all but barred to her, after completing secondary school, she took uncredited courses for graphic arts outside of university.[79]

Information about Ellen's prewar experiences comes primarily from oral history interviews with her elder sister Trudy, who lost contact with the family around 1941.[80] The letters and other material Ellen left behind after her death hardly touch on this period even as they document some of her wartime experiences.[81] Already before the war, Ellen may have wanted to follow in her father's footsteps and become a physician, but no account of her own remains that indicates this desire. Having completed secondary school in Germany before she and her family emigrated to the Netherlands, her academic ambitions were likely placed on hold in Amsterdam at least temporarily while she acclimated to her new home and a new language. No record describes the impact, but it could not have been easy, even for an intelligent and determined young woman. Initially, Ellen worked as an *au pair* in Amsterdam, while her parents, Julius and Dina, opened a boarding house (or, as Trudy called it, a sanatorium) for people who required special diets. This enabled Julius to practise as a physician and made good use of Dina's expertise as a dietitian.[82] The family lived in the boarding house, which was called *Huize Lairesse* (House Lairesse). The house still stands today in the elegant *Oud-Zuid* neighbourhood of Amsterdam.[83] It is possible that Ellen contributed to the family business in some way while also eventually studying at what Trudy described as 'medical laboratory school'.[84] The specifics of Ellen's higher education are unclear, but by 1941, she was working as a laboratory technician, almost certainly at one of the Jewish hospitals in Amsterdam due to the German racial laws.[85]

Whatever Ellen's education and career path had been, it was effectively halted by the German occupation of the Netherlands. By late 1940, new rules and limitations caught up with the thousands of Jews who had successfully escaped Germany for the Netherlands in the 1930s. In early 1941, all Jews were forced to register with the government. Jews were forbidden access to most public places, and their businesses were confiscated. According to Trudy, at some point, Ellen was arrested by the SS but was released within a few days.[86] She was fortunate, because in February 1941, several hundred Jews in Amsterdam were arrested and eventually deported to the Mauthausen concentration camp, where they were murdered.[87]

Ellen's and Eva's reticence over their experiences early in the war may indicate that their experiences of the war were most profoundly felt after the advent of the Final Solution in late 1941. Another reason may be that during this early period they were focused on merely existing amidst the chaos and turmoil of the time. Eva was similarly trying to make sense of her life at a time when few options for the future seemed realistic or even possible. Although she struggled to find a purpose, she continued to advance herself however she could. Ellen too was clearly determined to improve herself through education and a career despite the challenges imposed by flight from Germany and the occupation of the Netherlands. For both women, this was far from an aimless period. Although their actions were more tentative than they were for the other four women in this book, they still contributed to both pursuing STEM careers. There is no sense that either Ellen or Eva was destined for successful careers in STEM. Had their lives continued uninterrupted by the Second World War and the Holocaust, they would almost certainly have taken different directions than they ultimately did. The tragedy and persecution that they were to face during the Holocaust was designed to break them. But by their own accounts, their experiences of survival during the Final Solution inspired their future career paths. For all six women, however, life would have to become considerably worse before it got better.

The Beginning of the End

The rise of the Nazis in Germany and their growing strength and influence in Europe in the 1930s and early 1940s not only precipitated the start of the Second World War, it also interrupted the lives of many millions of people in Germany and Nazi-occupied and Axis European countries, including the six women whose lives are highlighted in this book. As racial minorities targeted by the Nazis, their lives were deemed inconsequential and dispensable. Nonetheless, as 1942 began, each in her own way had managed to survive arrests, aerial bombings, random violence, forced labour, deportations, and mass murder in their respective countries. Halina Neujahr, Halina Czajkowska, Ruth Pick, and Rita Levi-Montalcini all found purpose through education or work amidst the danger. As for Ellen Loeb and Eva Klein, little indicates that they had effectively found their future direction in life during this period. But they were not the only ones frustrated during this period. For the Nazis, the 'Jewish question' had become a Jewish problem. Despite their best efforts to make the expanding Reich *Judenfrei* (cleansed of Jews), the rapid acquisition of countries with large Jewish populations had gained millions of Jews and nothing they had done so far seemed to be reducing this problem in any significant way. A new solution had to be found.

Chapter Three

The Final Solution

> What we went through in these terrible times is difficult to put on paper. A book should be written about this.
>
> Ellen Loeb, MD, 1945[1]

As with the other women in this book, not all of Ellen's experiences have been recorded. Although each of them spoke and/or wrote to some extent about what they went through during the Second World War and the Holocaust, we can be sure that some aspects were perhaps either too painful to recall or were deemed inconsequential as compared to other experiences. What they and various documents did record is that their worst experiences of the war took place after 1941. In June of that year, barely three years after the joint Nazi and Soviet invasion of Poland, the Nazis invaded the Soviet Union. The two countries' unlikely alliance had been clouded from the beginning by the ideological differences of Fascism and Communism. Added to this was the fact that that the Russians were so-called 'Slavs' and were therefore abhorrent to the Nazis. Hitler and the Nazis also coveted the vast Soviet lands for more *Lebensraum*. Six months later, in December 1941, the United States entered the war after the bombing of the US Naval Base at Pearl Harbor, Hawaii by the Imperial Japanese Navy Air Service. Shifting alliances now meant that the Soviet Union and the United States were both on the Allied side of the war, fighting the Axis powers – Germany, Italy, Japan, Hungary, and others – alongside the United Kingdom and other nations.

Nazi Germany was also fighting another war, this one against the Jews. The regime had already deployed a variety of means to cleanse the Reich and all of Europe of Jews – forced emigration, mass murder, slave labour, ghettoisation, weaponised starvation and disease, just to name a few. The high command even considered forcibly sending Jews to Madagascar. But even after all of this, there were still around eleven million European Jews living in Europe. Conscious of the fact that millions more Jews lived in the Soviet Union, Operation Barbarossa – the name of the invasion of the Soviet Union – included mobile killing units, *Einsatzgruppen*, to murder Jews and others in unprecedented numbers.

By late 1941, it was clear to the Nazi high command that the Jewish question could 'only be solved by the biological eradication of the entire Jewry of Europe'.[2] The question remaining was how best to accomplish this on such a massive scale. On 20 January 1942, a conference of Nazi officials was held in Wannsee, Germany, to work out the details. German-American philosopher Hannah Arendt, who attended and reported on the 1962 trial of Nazi Adolf Eichmann, summarised the procedure of systematised genocide he helped to carry out:

> In country after country, the Jews had to register, were forced to wear the yellow badge for easy identification, were assembled and deported, the various shipments being directed to one or another of the extermination centers in the East, depending on their relative capacity at the moment; when a trainload of Jews arrived at a center, the strong among them were selected for work, often operating the extermination machinery, all others were immediately killed.[3]

While the racial laws had interrupted the lives of Ellen Loeb, Halina Neujahr, Halina Czajkowska, Ruth Pick, Rita Levi-Montalcini, and Eva Klein, the events of 1941 and after – including the Final Solution – were to have a far more devastating effect than they could have imagined.

Gateways to Auschwitz

In the Netherlands, Ellen Loeb and her parents were forced to wear the yellow Star of David starting in May 1942. The deportations began soon after, on 15 July. By mid-October, around 45,000 Dutch Jews had been deported to the Westerbork transit camp in the north of the Netherlands.

Time spent at Westerbork was usually short. It was a transit centre, after all – a processing and sorting centre for humans destined for annihilation. Most transported from Westerbork were sent to the extermination camps Auschwitz and Sobibor. There, those unfit for forced labour were sent to the gas chamber. The 'lucky' few were sent to the Theresienstadt Ghetto or the Bergen-Belsen concentration camp instead. Of the tens of thousands of Jews, Roma, and Sinti people sent to Westerbork, only a few thousand survived the Holocaust, earning it the dark sobriquet 'the gateway to Hell'.[4]

The Loebs were not among those on the first transports to Westerbork. A brief reprieve had been granted to them due to Julius's ill health. But deportation was all but inevitable. 'On the 8th of November 1942, our parents and I were aroused at 2:00 A.M. by the Gestapo and had to join all the other Jews to be transported to Westerbork,' Ellen later wrote to Trudy.[5] Once there, the Loeb family's goals were to stay together and avoid further transportation. They managed to do this in small increments. First, Julius was exempted from transport due to his status as a German military veteran and then due to ill health. If prisoners were ill, they could not be transported. Ellen avoided transport first by pretending to be younger than she was and then by feigning illness after intentionally taking twenty ephedrine tablets. This worked until April 1943, when Ellen and her mother Dina both became legitimately ill – Ellen with scarlet fever and Dina with a thumb infection. Julius soon followed, and all were spared from transport due to ill health. Once healthy again, Ellen intentionally gave herself diarrhoea by taking Castor oil.[6]

In addition to illness – real or induced – the Loebs prolonged their stay at Westerbork by making themselves useful. Julius's experience as a physician was valuable, Dina worked in the kitchen, and Ellen began working as a nurse at the primitive hospital. 'I knew absolutely nothing, but I acted as if I knew everything,' Ellen later wrote of her work. 'I just watched everybody wherever I could, and pretty soon I'd advanced to being a friend of the head nurse. After observing her for a short time, I learned how to take care of my patients.' With little medicine or equipment, Ellen worked her way through several specialisations, from women's internal medicine to surgery to the children's ward. Eventually, she was put in charge of paediatrics in June 1943. In this capacity, she practised the same kind of care for her young patients as she would later as a physician. Explaining how she and the other medical professionals prepared for the arrival of twenty sick children arriving from another concentration camp, she wrote:

> We tried to be ready for them. We prepared a room with big and little beds. We didn't have any linens, or at least we had very few. The only thing to give this room a friendlier appearance were some wonderful blue flowers we had been picking in the meadow outside that we put in water glasses all around.[7]

When the children arrived, there were forty-five of them rather than twenty. Many were infants and toddlers, 'all of them very, very ill. They were half naked or wrapped only in a thin blanket. Most of them were cried [sic] miserably and were totally exhausted and hungry.' Under the care of Ellen and the other paediatrics staff, all but five of the children were saved. Tragically, this meant that they were transported to their deaths in the gas chamber. 'It was wonderful and thankful work, but I knew that in the end they would be taken to Auschwitz,' she wrote.[8]

The Loebs lived in constant fear of being transported. Seven times they were put on the transport list and seven times they were spared.[9] In October 1943, Julius came down with pneumonia and had to be hospitalised, but Ellen and Dina were able to help care for him until his death on 2 November 1943. In the wake of Julius's death, the two women lost the protection his status as a doctor and a German war veteran had afforded them. They managed to avoid transport until February 1944, when they were deported to the Theresienstadt Ghetto.

* * *

Because of their work at the Jewish hospital in Prague, Ruth and Alfred Pick avoided deportation longer than most Jews in the Protectorate of Bohemia and Moravia. Between November 1941 and March 1945, nearly 74,000 Jews in the Protectorate were deported to the Theresienstadt Ghetto, which was in the eighteenth-century Fortress of Terezín, northwest of Prague.[10] In June 1940, the Nazis converted the fortress to use as a Gestapo prison and military base, before establishing it as a Jewish ghetto in late 1941. Ruth's fantasy that Theresienstadt might be a haven for Jews, especially compared to life in Nazi-occupied Prague, was an idea perpetuated by the Nazis. It was initially characterised as an 'old-age ghetto' for Jews age 65 and above and those who were wounded or had served with distinction in the First World War. Later, the Nazis called it a ghetto for 'privileged' Jews and even a 'model ghetto', which they would temporarily beautify to show to international aid organisations like the Red Cross. The reality was that the Theresienstadt Ghetto was a transit centre not unlike Westerbork, a 'gateway to death' where many Jews died of disease and hunger and even more were transported to extermination camps like Auschwitz and Treblinka, where they were murdered in the gas chambers.[11] According to the United States Holocaust Memorial Museum, 'Of the approximately 140,000 Jews transferred to Theresienstadt, nearly 90,000 were deported to points further east and almost certain death. Roughly 33,000 died in Theresienstadt itself.' The death rate in the ghetto was so high that in 1942, the Nazis built a crematorium just for the ghetto that could handle nearly 200 bodies a day.[12]

Ruth and Alfred were among the approximately 7,000 remaining Jews living in the Protectorate of Bohemia and Moravia when they were finally deported to Theresienstadt in July 1943.[13] Like all the others who were deported from Prague, they were forced to report to a central location where they went through processing that could last days. In addition to filling in forms and surrendering personal property and valuables, the SS – the so-called 'Protection Squadron' of Nazi Germany – performed interrogations that could include torture. While waiting for this ordeal to be completed, the prisoners lived in shacks with no toilets or floors that were guarded by SS guards and Czech police. Finally, they were brought to the central train station where they were loaded onto trains that took them to Theresienstadt. 'The boarding of the deportation trains took several hours,' explains one historian, 'and after a two- to three-hour trip they arrived at Bohušovice station, where they were received with shouts and beatings. There they were again checked (extra belongings were confiscated) before finally being marched to Terezín.'[14] Life inside the ghetto provided no relief. After arriving at Theresienstadt, Ruth and Alfred were separated into men's and women's living quarters – barracks and buildings crammed with triple bunks infested with bedbugs where disease and illness spread rapidly. Rations were sparse – imitation coffee, bread, and watery soup with a few potatoes – and had to be earned through work.[15]

* * *

As had been true for the Loebs in Westerbork, medical professionals were valuable prisoners in Theresienstadt and, in turn, medical work provided prisoners with some protection from the transports. The health system in Theresienstadt was extensive, and included a main hospital, auxiliary hospitals, a children's hospital, an infirmary, and an elderly care home.[16] Almost every medical specialisation was represented, including gynaecology, urology, dermatology, cardiology, infectious disease and psychiatric wards, and dozens of outpatient clinics.[17] All of this was necessary not because the Nazis wanted to save the lives of the sick and dying prisoners but because, as with the Warsaw Ghetto, they were afraid of the spread of disease for their own sakes.[18] Prisoner physicians, nurses, and other medical personnel were vital for handling hygiene issues like lice infestations, outbreaks of infectious diseases such as typhoid, tuberculosis, and scarlet fever, not to mention health issues related to age (both old and young), poor sanitation, and starvation. They were also expected to make their patients healthy enough for transport to the extermination camps.

In mid-1942, when the Theresienstadt health services were at their peak, there were 563 physicians and 1,583 nurses. But this was not enough to treat the number of sick and dying prisoners, nor were the facilities and supplies anywhere near sufficient. 'In the central hospital and various clinics, there were shortages of physicians, nurses, caregivers, and nursing assistants, and of medical devices and basic medical equipment such as bandages, sheets, towels, basins, urine containers, and so on,' writes Holocaust historian Margalit Shlain. 'Everything was filled with dirt and feces.'[19] Ruth discovered the latter soon after her arrival, when she was assigned to clean the infirmary. While scrubbing the floor, she realised it was covered in fleas. Shocked and disgusted, she was told by other prisoners who had been in the ghetto longer that the solution was to not clean so well.[20]

Ruth's anecdote reveals another fact of life about being a medical professional in the Theresienstadt Ghetto. According to historian Anna Hájková, 'women doctors were one of the groups who found it difficult to work in their profession in the ghetto.'[21] Although Ruth was a qualified physician, she had to start from the bottom and work her way up. After cleaning the infirmary, she was assigned to distribute clean laundry to the barracks. Eventually, she was fortunate to be given work as a physician. Her first medical job was as a doctor in one of the barracks, where she treated elderly German women suffering from nutritional oedema – severe swelling caused by malnutrition – without access to most medications. Later, she was assigned to treat tuberculosis patients, all of whom died because of the lack of necessary medical supplies.[22]

After Ellen and her mother arrived at Theresienstadt in February 1944, Ellen was given work as a nurse in the surgical department of the ghetto hospital. In the gender and social hierarchies of the medical system in the ghetto, nursing was women's work and so did not create the kind of difficulties for Ellen that Ruth encountered as a physician. Moreover, no training was necessary to work as a nurse in Theresienstadt, so desperate was the need.[23] It was in Theresienstadt that Ellen had what may have been her first encounter with her future career when was assigned to work with German doctors testing a new treatment for blood diseases, *heilgas*.[24] Because the treatment was led by German doctors, the medical personnel had access to medicines not otherwise available in the ghetto.[25]

* * *

Although Ruth's and Ellen's internment in the Theresienstadt Ghetto overlapped and they were both medical workers, it is probable that they never met. The sheer scope of the medical facilities at Theresienstadt and the fact that they worked in different departments makes it unlikely. After all, Ruth's husband Alfred also worked as a physician in the ghetto, but she was only able to see him occasionally. Ruth's and Ellen's fates were closer than they could have realised, however. Even though both were protected to an extent by their work as medical professionals, the end was near for the Theresienstadt Ghetto by the autumn of 1944. Ruth later recalled that she and other prisoners lived in constant fear of the transports that regularly thinned out the ghetto population. Their fear was based on the belief that they would be sent to forced labour camps. They did not know that those selected for transport were sent to Auschwitz and other death camps. 'They could not kill us fast enough,' Ruth said in retrospect.[26] She was right. Between 28 September and 28 October 1944, 'two thirds of the ghetto population – 18,402 people – were deported from Theresienstadt to the Auschwitz extermination camp'.[27] Both Ruth and Ellen were among them.

When Ellen's mother Dina was put on the list for transport, Ellen insisted on going with her. On 11 October 1944, the two women were loaded onto cattle wagons bound for Auschwitz-Birkenau, where they arrived the next day.[28] Less than two weeks later, on 22 October 1944, Ruth and Alfred Pick were also deported to Auschwitz.[29]

Deportations, Uprisings, and Defeats

By the fall of 1942, Halina Neujahr had been living, barely, in the Warsaw Ghetto for three years. She had been caught up in roundups multiple times and had managed to escape on several occasions. Before the implementation of the Final Solution, the roundups were typically for forced labour detachments. Forced or slave labour was an integral part of the Nazi war machine. The German economy depended on non-German slaves to keep ethnic Germans, the military, and the subjugated territories in sufficient condition to maintain morale and win the war. The lives of the slave labourers meant nothing to the Nazis, to whom their deaths were insignificant.[30] While some forced labour detachments in the Warsaw Ghetto were local, which meant that those rounded up could return at some point, as the war progressed, most forced labourers never returned.[31] Early on, Halina disconnected herself emotionally from the experience of being rounded up and sent on forced labour detachments. 'The psychology was such that the heavier the situation was, the least [sic] we thought about the peril ahead,' she later reflected. 'It was just to concentrate to survive the day, the hour, the minute.'[32]

The forced labour roundups also affected non-Jewish Poles outside of the Warsaw Ghetto. Throughout the war, no fewer than 1.5 million non-Jewish Poles, among them many teenagers like Halina Czajkowska, were transported to the German Reich to perform forced labour.[33] At age 15, Halina was still studying at the clandestine school in Warsaw, and observed the roundups, which she described in her memoirs:

> Two large lorries covered with tarpaulins would drive down a street, any street; when they stopped, the tarpaulins would open, and soldiers concealed inside

> would jump out. The two lorries would be stopped about 200 or 300 metres apart with no side streets between them. This action effectively trapped all pedestrians between the trucks [...]. The two groups of soldiers would then work toward each other, grabbing and pushing the trapped people against the walls. Anyone with German civilian documents was allowed to go, but all others, often close to a hundred people, were herded at gunpoint into the trucks and taken away. Usually they were sent to Germany. That was one way that the Germans furnished their slave labor.[34]

In the summer of 1942, the purpose of the roundups in the Warsaw Ghetto changed. No longer were Jews being sent to work as slave labourers. Instead, they were being deported to their deaths. 'The mass deportation, or "resettlement," of the Jews of Warsaw began on Wednesday, July 22, 1942 [...] and continued, with occasional short pauses, until September 12, 1942,' writes historian Yisrael Gutman. 'During those seven weeks, some 265,000 Jews were uprooted from Warsaw, transported to the Treblinka death camp, and murdered in the gas chambers.'[35] Toward the end of this period, known as *Grossaktion Warsaw*, Halina Neujahr was rounded up for deportation twice. Miraculously, she managed to escape both times. Her physical avoidance and emotional detachment during this time was accompanied by disbelief over the fate of Polish Jews.[36] But Halina Czajkowska and others were aware of the fate of Poland's Jews from Polish underground newspapers. 'The underground AK newspapers informed us that their destination was Auschwitz (Oswiecim) where mass murder took place in the gas chambers.'[37]

* * *

Having avoided the roundups, both Halina Neujahr and Halina Czajkowska were in Warsaw long enough to experience the uprisings against the Nazis carried out by Jewish and non-Jewish Poles. After *Grossaktion Warsaw* in mid-1942, it became clear to the diminished population of the Warsaw Ghetto what their fate was to be. It was only a matter of when. The Neujahr women were among only 40,000 to 50,000 Jews remaining in the ghetto in January 1943, when another 5,000 were rounded up and deported.[38] The three women barely escaped this event, which caught them unaware, as Halina later described.

> One very cold morning, the house in which we were then living, was surrounded by German soldiers shouting 'Raus, raus, alle raus!' [Out, out, everyone out!]. Our first impulse was to run for the shelter in the cellar, but the time was short and all the staircases were already blocked by German soldiers and Polish police. I managed to push my mother and my sister under a table, camouflaging it with two sacks of potatoes. Myself, I slipped between a wall and an open door leading from the hall to the kitchen. There were several towels hanging there, covering at least my head and shoulders. I heard heavy boots stamping through the apartment and voices of soldiers. Then I heard a womans [sic] terrified shriek and a shot. I felt all my blood rushing to my heart – it would burst any moment. Then I heard a voice, very close, '*heir is niemand mehr da*' [there is no one here anymore] and I heard heavy boots stamping away. I was sure that my mother was dead, but that was not so.[39]

The January deportation – or *aktion* – was just the prelude to the Nazis' plan to liquidate the Warsaw Ghetto, which began on 19 April 1943. Like most of the remaining Jews in the ghetto, Halina Neujahr, her mother, and her sister had moved into an underground bunker. When the Nazis marched into the ghetto, the Warsaw Ghetto Uprising began immediately, having been prepared for months. From the bunker, Halina could hear the uprising taking place above, as gunshots were exchanged and the Nazis brought in heavy artillery that shook the walls. This culmination of events brought a measure of comfort to 18-year-old Halina. 'In a way we were joyous, because for the first time – or at least, finally – we were masters of our own fate,' she explained. 'We were not afraid. We were there together, and we didn't think what would happen to us.'[40]

After around ten days in the bunker, Halina and her sister Sara went out during an apparent lull in the fighting to find food, water, and other supplies, leaving their mother behind in the bunker. They found the ghetto in flames and were immediately rounded up and taken to the infamous Warsaw *Umschlagplatz*, where those being deported were held before being loaded like cattle onto trains destined for concentration and extermination camps. 'We were packed into these freight cars. I think it must have been one hundred people per car. Many children with their mothers. And there was no sanitary arrangement, nothing. No water,' Halina recalled. After three or four days in these conditions, Halina and around 1,000 others arrived at a train station and were transported by trucks to the Majdanek concentration and extermination camp near Lublin, Poland.

* * *

Halina Czajkowska watched in horror as the Nazis retaliated against the Warsaw Ghetto Uprising. 'When it was all over and the fires died down, the Germans brought in bulldozers, and the entire area, where thousands of people had lived and fought so bravely, was levelled to a stone field.'[41] In two years, the population of the Warsaw Ghetto had gone from around 450,000 to nothing.[42]

As she witnessed the destruction of Warsaw's remaining Jewish population, Halina Czajkowska was closer than she realised to going through a similar experience as Halina Neujahr. Life had gotten progressively worse for non-Jewish Poles in Warsaw. In late 1942 and early 1943, mass public executions of non-Jewish Poles were increasing in the city in retaliation for Polish underground resistance activities.[43] Halina C. describes this unforgettably traumatic experience in her memoirs:

> They would do their usual *lapanki* [roundups], then move women and children to one side, holding them there while they executed the menfolk right before their eyes. Next day, the process was reversed; the women were shot. The bodies were left lying in the street, and people were forbidden to recover the remains of their loved ones. Eventually, a German truck would collect the murdered corpses.[44]

These actions only served to increase resistance activities among non-Jewish Poles, who saw in the annihilation of the Jewish population of Warsaw the possibility of their own fate.[45] As the Jewish population of the Warsaw Ghetto had done, the non-Jewish Polish citizens of Warsaw instigated an uprising against the Nazi oppressors on 1 August 1944, at the same time the Soviet army was advancing toward the city.[46] Halina and her mother hid in the basement of their apartment building while the German army retaliated against the poorly equipped insurgents. 'We were bombarded from above. As we had no antiaircraft weapons, the planes flew and did much damage with high explosives and firebombs,' Halina explained. 'The city started to burn. There were tanks on the streets, artillery, handguns, the railway weapons, hand grenades, mortars, and rocket projectors [...]. We knew that the battle was lost; this was a calculated slaughter of the civilian population.'[47]

Halina and her mother were relatively lucky in that they were not subject to the indiscriminate killing of non-combatant citizens – especially women and children – that characterised the first phase of the Nazi response to the Warsaw Uprising. In this early period, tens of thousands of innocent people were massacred by the Nazi SS. Tens of thousands more would be killed before the uprising was suppressed, but Halina managed to avoid this fate when the basement she was hiding in was raided on 10 August 1944. 'At the open door, we saw a machine gun facing us and shooting,' Halina explained. 'The bullets were ricocheting from the bricks above our heads; its aim was set to be lowered in the blink of an eye should we attempt to escape.'[48] Halina's mother was not with her at the time of the raid, having gone out to find food. As an unaccompanied teenager, Halina was especially vulnerable to sexual violence by the Nazi soldiers. When a soldier approached her and began to search her, an older woman bravely stood up to him at the risk of her own life. 'This woman, whose name I never learned,' wrote Halina, 'saved me from rape or a terrible death, which happened to so many others.'[49]

After a forced march that lasted around six hours, Halina and the other prisoners arrived at a former airport, where they were to be deported by train. There, she and her mother were reunited. Together, they boarded a train car with at least seventy-five others crammed inside, no water or food, and only two buckets for toilets. On the journey, the prisoners speculated about their destination. Soon enough, they knew without a doubt that they had arrived at Auschwitz-Birkenau.[50]

Selected for a Slow Death

Majdanek and Auschwitz-Birkenau were each a combination of concentration and extermination camp. Survivors of these camps often talk about their experiences of being sorted into lines on arrival. This was the selection process that determined who would live to become forced labourers and who would be sent to the gas chambers. The process was similar at both camps. At Auschwitz-Birkenau:

> families were divided after leaving the train cars and all the people were lined up in two columns. The men and older boys were in one column, and the women and children of both sexes in the other. Next, the people were led to the camp doctors and other camp functionaries conducting selection. They judged the people standing before them on sight and, sometimes eliciting a brief declaration as to their age and occupation, decided whether they would live or die. Age was one of the principal criteria for selection. As a rule, all children below 16 years of age (from 1944, below 14) and the elderly were sent to die.[51]

And, at Majdanek:

> Directly upon their arrival at Majdanek, the newcomers were stripped of all their personal belongings. Their hair was cut and sold from the camp to be used as material in industrial production. [...] The Jewish newcomers were subjected to preliminary selections. The SS-physicians generally sent the children, the elderly, and those the SS-men deemed 'unfit for labour' directly to the gas chambers.[52]

Halina Neujahr, Halina Czajkowska, Ellen Loeb, and Ruth Pick each experienced this grisly routine. Of the four women, Halina Neujahr was the first to face the selection process when she and her sister Sara arrived at Majdanek in early May 1943. After several days on the train, the prisoners were forced off and made to march through woods, where Halina feared that they would all be shot. Eventually, as dusk fell, she saw barracks and a watchtower in the distance. She later described what she experienced on arrival at the camp:

> Before the intake the next morning, after a night spent outside, children were separated from their mothers. Many mothers refused, fighting desperately the guards. Several were shot on the spot. Others remained hanging on the electrified barbed wire as they tried to reach their children, which had been placed in a special enclosure in the camp – a 'Kinderlager.' After [a] few days, all children were taken away in trucks and gassed.[53]

Halina and her sister were spared an immediate death in the gas chamber and instead selected for a slow death as forced labourers. The workday was long – twelve hours – and the work was brutal. 'Hard work in open fields or stone-pits: digging, carrying large concrete blocks, iron bars or simply stones,' Halina explained. Harsh punishments were delivered for the slightest infraction, such as daring to stop for a moment. Beatings were the norm, but Halina specifically recalled two of the worst punishments she experienced. The first was when she spilled the contents of the latrine she was carrying to empty, and the guard shoved her face into the excrement. The second was when a particularly feared female guard forced the prisoners to remove their shoes while they were walking to the work site. The shoes were never returned, and the women had to do their work from then on without them. What made this punishment worse was that the women were subject to daily selections, during which prisoners with injuries or illnesses were sent to the gas chamber. Without shoes, the women risked serious injury from walking over stones and broken glass with bare feet; injuries that could send them to their deaths. The gas chambers were never far away. Not only were all the slave labourers just one injury or illness away from being sent to their death in them, but they could also see and smell the victims' remains going up in smoke. As Halina recalled, 'The flames and the smoke from [the crematorium's] enormous chimney was [sic] always before our eyes and the sweetish smell of burnt human bodies in our nostrils.'[54]

Halina Czajkowska was next to face selection, arriving at Auschwitz-Birkenau in August 1944. To Halina C., her mother Emilia, and the other prisoners on the train, the word 'Auschwitz' meant one thing: death. What Halina experienced after stepping off the train did nothing to change that view.

> The Germans immediately began dividing people into groups, separating men from women and children. Suddenly, everything became very noisy; people were saying goodbye, some crying loudly. A woman parted from her husband could not control herself and became hysterical; shots rang out and both bodies fell. Silence followed that horror.[55]

Halina, Emilia, and the other women from the transport were marched off separately from the men and locked into buildings without windows or electricity. For several days, the women had no contact with their captors and were given no food and little water. 'Rumors were flying that we would be starved to death in this barn, like some Russian soldiers we had heard about. Or, they said, we would be burned alive.'[56] When they were finally let out, they were processed: stripped of all valuables and then of their clothes, forced to bathe in what was to them a mysterious dark brown liquid (in fact, a disinfectant), having their heads shaved, sent into showers, and given shabby dresses and wooden shoes to wear. Still wet from the shower, they were each allotted one small piece of bread and left to spend the night in a freezing cold outdoor space of the camp. Throughout this period, they were left to ponder their fate. Finally, in the morning, they learned that they were among the 'lucky' ones who were to be forced labourers for the Reich.[57]

* * *

Ellen Loeb and her mother Dina had created an incredibly strong bond in Westerbork, especially after Julius's death. As with Halina Neujahr and her sister Sara, and Halina Czajkowska and her mother Emilia, Ellen and Dina were determined not to be separated by the Nazis. But when their transport arrived at Auschwitz-Birkenau in October 1944, it seemed separation would be inevitable. With her mother, Ellen experienced the typical selection process, as she described soon after the war:

> When we arrived, everybody had to get out of the train. Your belongings stayed on the train. We never saw them again. Then we had to line up – men on one side, women on the other. The SS came and looked us over. Old people, mothers with children, and sickly looking women would all be sent to one side, and that meant the gas chamber.[58]

Despite her ordeal in Westerbork, Ellen was still strong enough and of the right age to be selected for forced labour. Fifty-year-old Dina, on the other hand, was selected for the gas chamber because her legs were swollen, possibly due to oedema caused by a heart condition. With immense courage, Dina objected to the Nazi medical officer evaluating her, denying that she was unfit for labour and saying defiantly, 'I'm going with my daughter. I can walk for six hours.'[59] It worked, and Dina was sent to the forced labour line with Ellen. The problem was now, as it had been for Halina Neujahr, to live in the terrible conditions of the camp barracks and not become ill or injured. 'If you lost weight you went into the gas chamber,' explained Ellen. 'You could not get sick because if you did, you would be killed. A physical exam meant only that you had to strip and the physician looked at you. Old, sickly, and thin, meager looking people went immediately to the gas chamber. The others went to work.'[60] Dina was nearly selected for death a second time when she was too slow to dress for transport to the labour camp. Again, she was spared, and after two weeks in Auschwitz-Birkenau, Ellen and Dina were forced onto cattle cars once again for a three-day journey to a forced labour camp in Austria, near the Mauthausen concentration camp.[61]

Arriving at Auschwitz-Birkenau just weeks after Ellen Loeb, Ruth Pick and her husband Alfred were also separated in the selection process, but not only because of gender. Alfred had made a pact with two of his male medical colleagues from Theresienstadt that they would stay together. At selection, Alfred was placed in one line while an older physician colleague with health problems was placed in another. Observing the pact, Alfred and his other colleague asked the Nazi officials if they could remain with their elderly colleague and were granted permission. Now unknowingly in line for the gas chamber, one of the men gave Alfred a piece of chocolate that he had somehow managed to save. 'My husband knew what a chocoholic I am,' explains Ruth. 'He went out of line to come over and give me the chocolate.' The consequence of this was that the Nazis beat Alfred while his two colleagues were led away to the gas chamber. When the beating was finished, Alfred was sent back to the line for forced labour, having narrowly avoided his colleagues' fate. Ruth was also selected for labour, unaware of how relatively fortunate she and her husband were to have been selected for a slow death, which at least gave them a chance to survive.[62]

* * *

By late 1944, the Nazis needed all the help they could get to keep the war machine going. Since all able-bodied men and boys in the Reich were serving in the military, forced labourers had become more essential than ever.

Although this imperative saved many Jews and others from an immediate death in the gas chamber, the Nazi plan to commit extermination through labour was as strong as ever. More than one thousand concentration camp subcamps and other facilities existed in Reich territories, in which millions of concentration camp prisoners, Jewish prisoners, prisoners of war, and civilians were forced to work for the Nazi cause.[63]

Ellen and her mother were sent to the women's subcamp of the Mauthausen concentration camp, Lenzing, where the prisoners were assigned to work as forced labourers in a rayon factory, which involved working with sulfuric acid with no protective equipment.[64] 'The men wore gas masks, but the Germans didn't give the women any,' Ellen explained. 'People were blind for days at a time because they didn't have any goggles. Others had stomach poisoning and were vomiting. Many could not eat. Many had wounds on their hands which immediately developed into purulent sores.'[65] Ellen managed to avoid this dangerous labour herself by again working as a nurse in Lenzing. But as the person treating these symptoms, she was acutely aware of the results of it.

Ellen's work did not exempt her from the conditions in the camp. On arriving in Lenzing, she and the other prisoners were given prison clothing infested with lice. With little food, the inmates resorted to eating whatever they could get a hold of, including flowers, discarded potato peels, and snails. Like all the prisoners, Ellen still had to do her work regardless not only of her exhaustion and hunger but also of other limitations. For her, the only problem greater than having too few instruments, supplies, and medication to help her treat her patients was the restriction on the number of sick patients she could have at one time.

> Of the 560 [women in the camp], I had 120 ill. Only forty were allowed. The others had to go to Mauthausen, and that meant gas, so I had to declare them fit for work if they were at all able to work, and that was one of the worst things. The women who were sick were too weak to send to a factory, but everybody knew they had to go if they wanted to live.[66]

Having left Auschwitz-Birkenau sometime after Ellen, Ruth was among around 200 women transported to Oederan, Germany, where she was interned in a subcamp of the Flossenbürg concentration camp in late October 1944.[67]

There, the women were forced to work in a munitions factory making anti-aircraft shells. Ruth was assigned to conduct Ph tests to determine the acidity levels of the chemicals used in making the munitions. Fortunately, not long after her arrival, the production of the shells became all but impossible due to shortages of raw material and malfunctioning machines. To keep the women busy, each day the Nazi guards forced them to pointlessly move scrap metal from one location to another and back again. Two inmates worked as physicians at the camp, but Ruth was loath to let the camp authorities know that she was a physician for fear of being transported to another camp. Though she and the other inmates were starving, by Ruth's estimation, conditions in the camp were not as bad as they might have been since they had access to water and were free of lice. Although Ruth did not officially work as a physician in Oederan, many of her fellow inmates knew she was a medical doctor and asked her for treatment. 'But nobody wanted to be sick,' Ruth later said. 'Because one was sick and they took her out and we never saw her again.'[68]

In the labour camps, staying alive was not only a matter of avoiding illness or pretending not to be ill. Although extermination through labour was the intent, the brutality of the guards ensured that many slave labourers were subjected to violent deaths. Halina Neujahr was all too aware of this in the Skarżysko-Kamienna slave labour camp in Poland, where she was transported from Majdanek in the summer of 1944. Working in the ammunition factory, Halina's job was to inspect cartridge shells passing by on a conveyor belt and remove those that did not pass quality control.

> With the long working hours, insufficient sleep, bad nutrition and general exhaustion, errors were not infrequent. The punishment was severe. Twenty raps on a poor woman's back with heavy sticks, held by two strong men, usually left her kidneys smashed up. Many died in terrible agony after such punishment. To this day, I don't understand how I was saved from such a fate.[69]

As if this were not bad enough, Halina N. acknowledged that there was another fate that was even worse. She felt fortunate that she was not one of the prisoners forced to manufacture explosives who had to handle picric acid – a highly toxic and combustible substance – without any protective gear. According to Halina, those who did this work 'became yellow all over, after only a few weeks, then changing into orange-coloured skeletons, shuffling their feet in slower and slower movements, before collapsing dead.'[70] Unsurprisingly, approximately 70 per cent of the Jewish slave labourers at Skarżysko-Kamienna died in the camp.[71] Halina's time in the camp was short and she was soon transferred to another labour camp in Częstochowa, Poland, where conditions were marginally better. In late December 1944, she was transferred again, this time to the Ravensbrück concentration camp, where her forced labour was perhaps more painful emotionally than physically since it involved unloading and loading from trains boxes of looted Jewish belongings, which Halina observed included 'abandoned toys and family photographs of annihilated families'.[72]

In the Bergen-Belsen concentration camp, where Halina Czajkowska and her mother were transported in November 1944 from Auschwitz-Birkenau, the two women managed to be assigned to relatively better labour in the kitchens, peeling and shredding potatoes and other vegetables near a warm oven. But no labour was easy labour, and both women's hands soon became as shredded as the vegetables, causing them to be reassigned.[73] Slave labour became the least of Halina C.'s concerns in early January 1945, just before her 18th birthday, when she became very ill and was sent to the camp *revier*, a pseudo medical centre. Halina's experience as a patient provides a complementary perspective to the experiences of Ellen and Ruth as medical professionals. While the latter were preoccupied with the lack of medicines and equipment needed to treat their patients and with helping them avoid transport, the patients had other things to fear. They knew that that the *revier* was often a last stop before the gas chamber. Worse, it was a place of medical experimentation, where the prisoners were the Nazis' unwilling laboratory rats. But even if a prisoner managed to avoid becoming a victim of a medical experiment or being sent to the gas chamber, the conditions of the *revier* and the treatments performed were horrific.

In the *revier*, Halina C. was fortunate to have a caring female Polish physician – a fellow prisoner – treating her. She diagnosed Halina with fluid in her right lung that was increasing and could lead to death. Under normal conditions, the fluid could be drained, but this procedure was impossible without the proper instruments. With few other options, the physician suggested a radical course of action: an injection of calcium that would either cure Halina or cause a fatal allergic reaction. Halina was willing to take the risk and was given the injection that she later described felt 'like a fire or molten metal coursing through my veins. Heat flooded my head and the bottom of my feet simultaneously.'[74] Either despite or because of the procedure, her condition slowly improved. In the meantime, she had to recover in the *revier*, a place hardly conducive to healing since patients with various types of ailments slept two to a bed. As Halina described:

> Women with all kinds of ailments were put together. On the bunk directly below me, lay a woman with advanced tuberculosis who hemorrhaged when she coughed. She finally died with a terrible gush of blood. In a lower bunk about two feet away, was a nice young Russian woman soldier dying from the deadly *Bauch*-typhus; she had looked strong, but the high temperature and bloody diarrhea soon killed her too.[75]

While Halina was recovering in the *revier*, typhus and typhoid epidemics swept through Bergen-Belsen just as the camp population increased exponentially due to the evacuation transports arriving from other camps. Among those transported to Bergen-Belsen in late winter 1945 were Halina Neujahr and her sister. On the verge of losing the war, the Nazis did little to contain the epidemics and worsening conditions in the camp, which now housed around 60,000 starving inmates. Thousands died in the overcrowded camp in the final months of the Second World War. As they did, their corpses were thrown into mass graves or were piled on top of one another, creating mountains of bodies around which those who had not yet died but who looked like walking skeletons waited for their turn to come.

* * *

Death was all but inevitable in Nazi extermination, concentration, and forced labour camps. The only variable was the speed at which a prisoner would die. Some died relatively quickly in the gas chamber (though not quickly enough to prevent their suffering). Others died more slowly in concentration and forced labour camps as the result of labour, illness, starvation, and violence.

In labour camps alone, it is estimated that around 2.7 million people died or were murdered.[76]

Surviving inevitable death was more a matter of luck than anything. It is too easy to suggest that factors like determination and resilience alone kept people alive since no one was immune to violent punishment, illness, injury, starvation, and random selections. Ellen, Ruth, Halina Neujahr, and Halina Czajkowska all recognised that their own survival came down to chance. However, their individual stories also demonstrate that *taking* chances also played an important role in their survival. Risk-taking – such as escaping roundups, submitting to dangerous medical treatments, feigning both illness and health, and refusing to be separated – contributed to saving their own and others' lives at critical moments, enabling them to live to see another day. The same is true of Rita and Eva, though under very different circumstances.

Surviving on the Outside

Experiencing the Second World War and the Holocaust outside of concentration and labour camps was different, but it was still survival. Rita's and Eva's experiences demonstrate that surviving in Axis countries was as much about chance and taking risks as life in the camps, only with different dangers. Starting in 1943, these dangers increased in Hungary and Italy as alliances began to disintegrate, and the Nazis sought to take control of what they considered vassal states.

Until mid-1942, Rita lived a relatively peaceful existence in Turin, not to mention a productive one. Aware of the progress of the war and conscious of the anti-Jewish rhetoric in Italy, she conducted her experiments on chick embryos in her improvised home laboratory. Her self-appointed mission was to repeat experiments conducted by German embryologist Dr Viktor Hamburger of Washington University in St Louis, Missouri. She was intrigued – if not convinced – by Hamburger's arguments about embryonic nerve development published in a 1934 article which she read for the first time in 1940.[77] As Rita later described:

> Viktor was the first to develop the field of chick embryology. He started to inquire how target organs or limbs act on the development of nerve centers,

> which send nerves in to the periphery [limbs]. In 1934 he found that if you take away limb buds from an embryo, motor and sensory nerve cells, which should innervate the limbs, do not develop from the spinal cord. If you attach an additional limb they enlarge. His interpretation was in the spirit of experimental embryology. The failure of the nerve cells to develop in an embryo deprived of the limb buds was, in his mind, in line with the concept of embryonic 'induction.'[78]

Hamburger's theory was inspired by his former mentor, German embryologist Hans Spemann (1869–1941), who won the Nobel Prize in Physiology or Medicine in 1935 for the discovery of the 'organiser effect' in embryonic development, also known as embryonic induction.[79] Rita seemed to regard Spemann's concept with contempt, writing later, '*Induction* is a mystical term that was meaningless to me. No one knew what this vague word meant, but it pleased embryologists at the time. But it was a term I didn't accept.'[80] If her opinion was informed by Spemann's reported Nazi sympathies, she gave no indication of it.[81] Although unimpressed with induction, Rita was impressed by Hamburger's style and rigorous analysis, which she felt were lacking in other researchers' articles. In an article written in honour of Hamburger, published in 1981, Rita credited his article with starting her on the path to discovering NGF. It 'not only spurred my curiosity,' she wrote, but more importantly, 'suggested exploration of a problem that I could investigate in my bedroom-laboratory without having to resort to pieces of equipment [...] which were no longer available to me.'[82]

Having found her purpose, Rita dedicated herself to it starting in the spring of 1941 'in the calm of a minuscule laboratory not unlike a convent cell' in Turin. This cloistered existence did not last long. In late summer, she invited her former teacher, Giuseppe Levi, who had shared Hamburger's article with her in the first place, to collaborate with her on her research, and he accepted. For the next year, her bedroom-laboratory was filled with Levi's 'imperious voice' and his 'loyal pupils' whose presence meant that 'the madmen and criminals who were running our country' were discussed almost as much as the study of chick embryos. As Rita had intuited, what she and Levi found in their research was different from Hamburger's interpretation. According to Rita, the death of nerve cells 'appeared to be caused by the absence of a trophic factor and not, as Hamburger had hypothesized, by an inductive one belonging to the category of those known as "organizers".'[83] Their findings 'signalled the beginning of a new chapter in neurobiology', writes one neurophysiologist in retrospect.[84] But war does not pause for great scientific discoveries, and Rita's and Levi's findings were at risk, existentially and intellectually.

To make their findings known, they had to publish them. This was impossible to do in an Italian journal due to the country's racial laws prohibiting non-Aryans from publishing. Making matters more difficult, in the summer of 1942, the war came to Turin in the form of Allied aerial bombing of northern Italy. As much as Rita supported the cause of the Allies, she did not wish to become a martyr to it. Second only to her life and the lives of her family and Levi was the protection of her scientific research.

> Almost every night, the lugubrious whine of sirens, warning of British planes overhead, forced us to go down into the basement in spite of the risk – which became tragic reality for hundreds of people – of being buried under the ruins of bombed buildings. Every time the alarm sounded, I would carry down to the precarious safety of the cellars the Zeiss binocular microscope and my most precious silver-stained embryonic sections. These vigils usually dragged on for hours, amid the murmurs of women praying, until the sirens announced that the danger had, for the time being, come to an end.[85]

Acutely aware of the dangers of remaining in Turin, she and her family evacuated to a small house outside of the city, where she set up a new laboratory and continued her research in slightly less disruptive surroundings. Levi had also left Turin for safer ground and frequently came to work with Rita in the new laboratory. They even managed to have their findings published in a Belgian journal in 1942 and 1943 and in a Vatican publication in 1944.

In July 1943, there was a brief glimmer of hope for Italy when Italian Prime Minister and Fascist dictator Benito Mussolini was forced to resign and go into exile. Popular opinion in Italy was against the Nazis, and the new prime minister, Pietro Badoglio, agreed to an armistice with the Allied powers. But these developments provoked the wrath of Hitler and the Nazis, who moved to gain control of Mussolini and occupy Italy. In early September, the Italian royal family and the prime minister fled Rome for safety. With the Nazis occupying Turin and other Italian cities, Jewish Italians like Rita and her family were at a greater risk than ever. Mussolini had consistently refused to deport Italian citizens who were Jewish to Nazi concentration camps. Now, with the Nazis in charge, the roundups and transports began, mainly to Auschwitz-Birkenau. In early fall 1943, false identity papers in hand, Rita and her family attempted to flee to Switzerland. Only Rita's older sister Nina and her family were successful. Rita, her twin sister Paola, mother, brother and his family managed to get only as far as Nazi-occupied Florence. The three women were given a room by a non-Jewish woman on the condition that they assured her they were not Jewish. This was due not to antisemitism, Rita later explained, but rather to fear of Nazi reprisals for sheltering Jews. According to Rita, their host was aware from the beginning that her guests were Jewish.[86]

Unable to conduct her research, Rita spent some of her time filling out fake identification cards for others who had fled to Florence. Among those who escaped the Nazis and found refuge in Florence was Giuseppe Levi. He and Rita spent time walking together and discussing histology, no doubt frustrated at their inability to continue their work. The months must have passed slowly as they followed the progression of the war on the radio and suffered the privations of it each day. The Allied campaign to liberate Italy was underway. In early September 1943, Allied troops landed in southern Italy and began advancing northward, though not without considerable resistance by German forces, especially as the US, British, and Canadian troops got nearer to Rome. Securing the capital took most of the first half of 1944, but Rome was finally under Allied control by 4 June 1944, allowing for a somewhat faster advance northward towards Florence. In her memoirs, Rita recalls how on 3 August, a state of emergency was declared in Florence, as the German troops struggled to defend the city from the Allies. Movement during the day was curtailed, and a nighttime curfew was imposed. Supplies of water, bread, and electricity ceased. In the early hours of 4 August, Rita was awakened by 'the ominous sound of falling bridges'.[87] The destruction of Florentine bridges across the Arno river had been ordered by the German commander to slow the advance of Allied troops and allow the German forces to escape. Only the medieval Ponte Vecchio bridge remained.[88]

As the Allies advanced and the German troops retreated, chaos reigned in Florentine streets after the underground resistance group the Tuscan Committee of National Liberation organised an insurrection to undermine the last vestiges of German control. Beginning on 11 August, pockets of fighting between members of the resistance and the last of the German troops lasted until 1 September, when the battle for Florence finally ended. Rita recalled the danger and the exhilaration she felt during this period:

> From the roofs, snipers in the pay of the Nazis and Fascists were constantly firing at passersby. Even though the war was going on outside our windows (it seemed as if the battlefront ran right down the street where we lived, so frequent was the firing of rifle and machine gun which hit civilians and military alike), we were breathing an exhilarating air of freedom and still went out into the streets heedless of the dangers. We also risked stepping on one of the mines that the Germans had scattered about and which caused many deaths.[89]

On 2 September 1944, Rita watched as 'the British marched silently along the streets packed with onlookers'. The relief of liberation became even more palpable when she saw a bus adorned with a Star of David, 'now no longer an object of derision'.[90] Despite the threats Rita faced in Florence, she and other Jewish Italians had made the right choice to escape from Turin when the Nazis took over. Of the 4,733 Jews deported from northern Italy to Auschwitz-Birkenau, only 314 survived.[91]

* * *

In Hungary, the antisemitic head of state Miklós Horthy had resisted Hitler's demands regarding the deportation of Hungarian Jews, choosing instead to maintain his autonomy, not least where the Jewish question was concerned. As a result, Jewish Hungarians like Eva and her family lived in relative safety compared to Jews in Nazi-occupied territories like Poland and Bohemia and Moravia.[92] That is, until 19 March 1944, when the Nazis – frustrated with their ally's actions – marched into and occupied Hungary. Eva was walking in Budapest the day the Nazis occupied the city. Years later, she recalled a strange silence in the city. No resistance or fighting, just silence.[93]

With Adolf Eichmann now in charge of the Jewish question in Hungary, the approximately 800,000 Jews living in the country – including both Hungarian Jews and Jews from other countries who had fled to Hungary – suddenly found themselves facing the full wrath of the Nazis' racial policies. The process was familiar. All Jews were required to wear the yellow star, making them easily identifiable. Ghettos were established throughout the country, starting in the provinces. The plan was to gradually deport all of Hungary's Jews to Auschwitz-Birkenau from six geographical zones, of which Budapest was the last.[94] The Nazis' implementation of the Final Solution in Hungary was remarkable for the speed with which it was carried out. On 15 May 1944, mass deportations to Auschwitz-Birkenau began. Just eight weeks later, when Horthy halted the transports, more than half of the Jews in Hungary had been deported.[95] According to historian Randolph L. Braham:

> As a consequence of the occupation, the Hungarian Jewish community, which had survived the first four and a half years of the war relatively intact, was subjected to the most ruthless and concentrated destruction process of the Holocaust. The drive against the Hungarian Jews took place on the very eve of Allied victory, when the grisly details of the Final Solution were already known to world leaders, including those of Hungarian and world Jewry. Informed about the barbarity and speed with which the Hungarian Jews were liquidated, Winston Churchill concluded that this was 'probably the greatest and most horrible crime ever committed in the history of the world.'[96]

Encouraged by her family and friends, Eva briefly left Budapest and worked under a false identity thanks to a Christian friend, Clara Bayer (1924–2011), who gave her copies of her birth certificate and other papers. By the summer of 1944, she was back in Budapest, around the same time the deportations were stopped. Despite this turn of events, the danger was still great. Hearing rumours that working in the war industry provided some measure of protection, Eva applied for and was hired to work in a state brick factory located at the city limits. Soon, Eva, her sister, and several other Jewish women were not only working at the factory but also living there under the protection of the management. When the brick factory was damaged by a fire, she and the others were moved to a textile factory, where they had similar arrangements. Despite the precarity of their circumstances, Eva and her companions lived a relatively comfortable existence. The mother of one of the young Jewish women cooked for the group, and they spent their free time after work 'reading and having a good time', Eva later recalled. Everything changed after what she described as the fatal day of 15 October 1944, when Horthy was deposed by the Nazis and replaced with the Fascist Ferenc Szálasi as the head of a Nazi puppet state.[97]

Szálasi's leadership was fatal for Jews in Hungary not merely because he was happy to commit atrocities on behalf of the Nazis but also because he personally and zealously believed that Hungary must be cleansed of Jews.[98] In the 1930s, he had founded the far-right Arrow Cross Party on this and other extremist principles, and when he came to power in October 1944, he filled all positions of power with members of this party. Soon after taking over, the new government targeted Jews in Budapest, which had largely been spared the deportations of the spring and early summer and so was where most of Hungary's remaining Jews lived. Deportations were resumed, with tens of thousands of Jews sent to Nazi camps either by train or via forced marches. Two Jewish ghettos were established in Budapest, where tens of thousands more were imprisoned under horrific conditions. Jews who escaped the net were hunted down by Arrow Cross guards. 'Those captured were either shot and thrown into the Danube [river, in Budapest] or taken to the ghetto,' explains one historian.[99] Nearly 80,000 Jews alone met the former fate.[100]

It is no wonder Eva considered that she was only truly hiding after 15 October 1944. First, she moved into a textile dye factory that was ostensibly abandoned but which became a hiding place for Eva, her then-boyfriend – a rabbi – and several other Jews. Then, she moved into the histology department of the medical school, where she had access to the library and gained a fake identification card that allowed her to occasionally go outside. Soon, her parents, sister, sister-in-law, and others joined her in hiding at the medical school. While all her needs were met – food, shelter, safety, the comfort of family, and relative freedom of movement – emotionally, Eva was numb. 'I think I was living all the time in a dream,' she later recalled. 'Both the bad things and before. I had no future. There is [sic] nothing. I was even not afraid of what will happen.' Even in later life, she was self-deprecating. In her interview with the USC Shoah Foundation in 1997, when she was 72 years old, she described herself in her youth as 'not so clever and very passive'.[101] But this characterisation stands in sharp contrast to her actions in late December 1944.

On 21 December 1944, Eva's sister-in-law was walking outside their hiding place using a fake identification card when she met someone who shared news of her husband, Eva's brother, who was imprisoned in a forced labour camp in Balf, western Hungary. The contact gave her information about how to enter the camp and free him. Soon after, Eva obtained false papers for her brother, and the two women set out to the location of the camp on foot and by hitchhiking. On arrival, they encountered prisoners of the camp arriving back from work duty and stayed two nights among them, hoping that they could help find her brother. Eventually, the women were smuggled into the camp where they found him, and the three managed with the help of their false papers to leave the camp. Faced with many dangers, they returned to Budapest and their hiding place at the medical school in Budapest on Christmas Eve 1944, where they were greeted by a Christmas tree and festive food prepared by János Szirmai (1925–2014), the Christian friend helping to hide them. No sooner had they arrived than one threat was replaced by another as the Soviet army advanced into Hungary.[102]

Liberation

'Liberation is not deliverance,' French novelist Victor Hugo wrote in the nineteenth-century novel, *Les Misérables*.[103] Few knew the truth of this better than those liberated by Allied forces in 1944 and 1945. Feelings of relief would have been accompanied by a host of other emotions, not least fear, grief, anger, and confusion. What was left? Who was left? What happens now? Illnesses and starvation did not suddenly disappear upon liberation. Many died in the days, weeks, and months following liberation due to the irreversible effects of starvation, forced labour, and disease, or even eating too much too quickly. For women, the threat and reality of rape did not disappear upon liberation. In some cases, they were more likely to be sexually assaulted by the soldiers liberating them than they had been by those imprisoning and persecuting them.[104] Liberation from the Nazis was not deliverance from danger, violence, and death.

Rita's response to these ongoing threats was to volunteer her services as a physician. Soon after the British troops marched into Florence on 2 September 1944, she was assigned to work in a makeshift emergency field hospital, where truckloads of rural inhabitants arrived each day for medical attention. Having spent much of the war relatively sheltered, Rita was shocked by what she saw.

> The consequences of malnutrition and of the cold were particularly evident in babies and old people. I had never in the course of my brief experience as a practicing physician – before the racial laws prevented me from frequenting hospital wards – witnessed a sight so painful. Many of the newborn children arrived in a state of extreme dehydration. At dawn I would take them to the pediatric clinic, only to be an impotent witness to the extinguishing of the lives of those small creatures, by then already on the threshold of death.[105]

The situation in Budapest in early 1945 can hardly be described as 'liberation'. On 26 December 1944, the Soviet army surrounded the city and subsequently laid siege to it. Before the battle for Budapest ended with a Soviet victory on 13 February 1945, it took a devastating toll on the civilian population. Among the around 38,000 civilians who died during this period, nearly half were Jews, most of whom were murdered by the Hungarian Arrow Cross Party. Higher than the civilian death count is the rape count during the siege and subsequent Soviet occupation, which is estimated at around 50,000.[106] Eva and her family and friends in hiding were not protected from these dangers. On at least one occasion, a group of Soviet soldiers entered the university cellar where Eva and the others were hiding and 'invited' the occupants to drink large amounts of vodka with them. As Eva later explained, refusing this invitation was as dangerous as accepting it. On the one hand, refusing angered the soldiers. On the other hand, the drunker the soldiers got, the more violent they became. To protect her, Eva was locked her in another room but, for some reason, her sister-in-law remained with the soldiers, who attacked and attempted to rape her. Risking his own life, her husband (Eva's brother) ran out into the street and found a senior Soviet officer who returned with him to the cellar and put a stop to the violence.[107]

As the Allied forces made their way through countries like Italy and Hungary in late 1944 and early 1945, life in the Nazi concentration camps was going from bad to worse. During this period, Nazi officials began evacuating concentration and labour camps that were in the path of the Allies, often forcing the inmates to travel on foot for hundreds of miles to camps further inside the diminishing German Reich. Thousands of inmates died or were murdered on these death marches, while those who survived ended their journeys at already overcrowded camps, where contagious diseases were decimating the populations. One of these camps was the Bergen-Belsen concentration camp in northern Germany, where Halina Czajkowska arrived in November 1944 and Halina Neujahr in March 1945. Not long before Halina Neujahr arrived at Bergen-Belsen, an epidemic of severe diarrhoea known as *Durchfall* took hold of the prisoners in the camp. Halina Czajkowska later explained: 'A most tragic sight, these people with *Durchfall* were often dragged to the corridor, where they lay in their wet rags until they died. Every morning the number of dead was larger. You woke up with a corpse beside you or resting on your legs.'[108] This and outbreaks of typhus and tuberculosis, among others, claimed the lives of tens of thousands of inmates in the final months of the war and even after liberation. When British and Canadian troops liberated Bergen-Belsen on 15 April 1945, they were confronted with little more than corpses, either dead or barely clinging to life.[109] Among the latter were Halina Czajkowska and her mother, and Halina Neujahr and her sister. Had Bergen-Belsen been liberated any later, these women would likely have perished along with so many others.

The day before Bergen-Belsen was liberated, the Nazis evacuated the Oederan forced labour camp in Bavaria in anticipation of the Allied advance. Ruth and the other inmates were marched under armed guard into open cattle cars. The deterioration of the Nazi war machine was evidenced by the youth of the guards pointing guns at them. 'Little boys. Teenagers. Less than teenagers,' Ruth explained. Nonetheless, Ruth believed that the destination of the transport was Mauthausen concentration camp, where she and the other prisoners were to be gassed. As luck would have it, their train headed towards Theresienstadt instead. For one week, she and the other inmates stood in the crowded open train cars until suddenly, not far from Theresienstadt, the train stopped, and the young guards disappeared. The prisoners walked the rest of the way to Theresienstadt, where thousands of other prisoners were also arriving. There, they were given medical attention, first by other prisoners, and then in early May, by the International Red Cross. Shortly after, the Nazi camp commandant and SS guards fled, and on 11 May 1945, Theresienstadt and its 30,000 prisoners were liberated by the Soviet army.[110]

Ellen was treating patients affected by an outbreak of nutritional oedema when liberation day came for the Lenzing subcamp of Mauthausen concentration camp on 5 May 1945. In her letter to her sister, Trudy, Ellen wrote:

> Now came the big moment. We were freed. The plan had been to shoot us or gas us all, but I heard that the manager of the factory took his car and met the American Army and told them of the women in his factory. This influenced the high command to get to us two days earlier than had been planned. On Friday, the 5th of May, at 6:00 P.M. we heard the alarm. After [a] silence of five minutes, we knew we were free. The SS ran away, and on Saturday we were given over to the Americans. You cannot imagine what that meant. Most of us were at the end of our strength. A few more days, and we would not have made it.[111]

The war in Europe officially ended on 8 May 1945. Hitler had died of a self-inflicted gunshot wound on 30 April. With the liberation of the Nazi concentration camps, the Holocaust was also effectively over. And yet, the death count increased for many months and even years afterwards, as survivors succumbed to the effects of disease, starvation, and the psychological trauma of their experiences. Lives, homes, families, and careers all had to be rebuilt. Surviving the Holocaust and the Second World War had been one thing. Ongoing survival – deliverance – was another matter entirely.

Chapter Four

Afterlives

> Hunger and trauma had left their marks. As the studies began, they seemed extremely difficult, much more so than we had expected. We all experienced memory troubles, which at first each of us suffered in silence, feeling that something was wrong but not knowing what to do about it.
>
> Halina Czajkowska Robinson, 2020[1]

Soon after the liberation of Nazi concentration camps, images of the mountains of corpses and emaciated survivors that greeted the liberating troops were circulated in newspapers, magazines, and newsreels around the world. It had been years since Trudy Shakno, living with her family in Dallas, Texas, had heard any news of or from her sister, Ellen Loeb, and their parents. Seeing the 'atrocity photos' of the concentration camps would have given her little hope that they had survived. Those who survived the camps were equally uncertain about the fate of their loved ones. Ruth lost track of her husband, Alfred, following the selection at Auschwitz-Birkenau, when he was beaten by Nazi guards for getting out of line to give her a piece of chocolate. Halina Neujahr and her sister Sara managed to stay together since their arrest in the Warsaw Ghetto in April 1943. But the fate of their mother was unclear. Halina Czajkowska and her mother had also stayed together but had no knowledge of the fate of Halina's father.

For many survivors, life ended during the Holocaust even if the physical body never expired. Survival in the aftermath was far from guaranteed. Halina Neujahr and Halina Czajkowska both spent many months in hospitals receiving lifesaving treatment for the devastating effects of their time in concentration and labour camps. In nearly every location where survivors were given medical treatment and refuge following the end of the war, there are graves of those who perished in the days, weeks, and even years after liberation because of their physical and psychological traumas.

Establishing new lives was no more certain than living another day. Halina Czajkowska described the first few years of her life after liberation as purgatory. Rita considered the two years after liberation an extension of 'the difficult years' of her life. The steely determination that characterised her during the war suddenly left her in the aftermath. In the camps, Ellen's medical work had contributed to saving lives, including her own, and given her purpose. After liberation, her lack of formal education meant that her medical career was over. In Hungary and Czechoslovakia, Eva and Ruth were each faced with the reality that the end of one repressive regime was the beginning of another.

Yet the aftermath was also a time of rebirth. Though by then young adults, both Halina Neujahr and Halina Czajkowska returned to school to complete their basic education – the first step on the path to their future careers. Crises became important turning points. Ellen's crisis made her more determined to overcome the obstacles in front of her and become a physician, while Rita's involved her abandoning medical practice entirely in favour of research. After years of self-doubt and uncertainty, Eva found her calling in life amidst the chaos of the postwar period. She and Ruth also managed to escape their homelands before the fall of the Iron Curtain – with much of Eastern Europe (controlled by the Soviet Union) on one side and the 'West' – Western Europe and the United States – on the other.

In many ways, the Second World War and the Holocaust had dramatically changed each woman's lives for the worse. But thanks in no small part to luck, they were granted afterlives – life after death of a sort – in which their actions and decisions contributed to making their experiences of persecution and survival integral elements of their careers in, and contributions to, science and medicine.

Surviving Liberation

Not long before the liberation of the Bergen-Belsen concentration camp, 20-year-old Halina Neujahr was helping to carry the body of a 13-year-old girl who had died overnight to a tent where there was an ever-growing pile of corpses. Weak and ill herself, Halina lost consciousness. When she came to later in the day, she discovered that she was lying among the corpses in the tent, left for dead. Somehow, she managed to crawl out of the tent and return to her barrack.[2] She was lucky. Following the liberation of the camp on 15 April 1945, the liberating troops struggled to tell the difference between the living and the dead. British medical assistant William Arthur Wood described what he saw when he entered the camp a few days after liberation:

> And then on the left-hand side there were the huts. And of course, outside the huts were piles and piles of dead bodies, and living ones, and you didn't know which were which. In the huts themselves, equally, you didn't know who was dead and who was alive unless they made, you know, some movement you could see, because the dead and the living were all together. They hadn't the energy to take the dead out and there were so many piled outside, as I say, that it was hard to see, to pick out the dead from the living.[3]

In their barrack at Bergen-Belsen, Halina Czajkowska and her mother Emilia were among those offered a feast on liberation day, courtesy of the British soldiers. 'When the container was opened, a fantastic aroma wafted over us,' Halina remembered. 'It was a thick stew, rich in meat, fat, and vegetables. Many could not resist and quickly reached for it.' But Emilia knew better.

> She warned us that it was too rich and could kill us after such long deprivations. Some heeded her advice; others ignored it. Of the nearly 300 women [in the barrack], only about 150 were left alive the following day. The rich food killed most of those who ate it; their intestinal tracts could not handle the sudden meal.[4]

The liberating soldiers had meant well. They and even many medical professionals at the time were unaware of what is known today as refeeding syndrome, which occurs when individuals who have been subject to extreme malnutrition are reintroduced to food too rapidly. Understanding of the syndrome was hastened by what the liberators encountered in the camps when they gave the former prisoners food too quickly. Unfortunately, many died in the process, perhaps as many as a quarter of the approximately 14,000 who died after liberation succumbed to refeeding syndrome.[5]

But death post-liberation came in many forms. Shortly after liberation, Emilia Czajkowska became severely ill with *Durchfall*. Already weak from starvation and an earlier epidemic of typhus, her chances of survival were slim. 'Liberation notwithstanding, we had no one to call on for medical attention,' Halina Czajkowska wrote in her memoirs. 'Everybody needed help to some degree, and there simply were not enough medics available.'[6] As much as the liberators wanted to help the nearly 60,000 prisoners, they were completely unprepared for what they found at Bergen-Belsen. 'We'd been trained for war wounded. We were used to terrible wounds,' Wood recalled. 'But I'm afraid when we got to Belsen, we hadn't been trained for this… it was so different from anything we'd seen.'[7] Other problems were a lack of medical personnel and appropriate facilities for treating patients. The war had not ended, so most medical units were on or near the front lines. 'There was very, very little in the way of medication or equipment,' recalled Dr Laurence Wand, who was a medical student at the time. 'But then, even if there had been, the problem was one couldn't have used it because the problem was too overwhelming.'[8]

Communication was also an issue. After the other former prisoners had been removed from their barracks, 17-year-old Halina C. and her mother, who could barely move, remained behind. Having managed to stay together throughout the war, Halina's fear of being separated from Emilia was triggered by the Allied medical personnel talking to her in incomprehensible languages. Desperately clinging to her mother, she refused to be separated from her. Finally, when a Polish speaker addressed her, she began to cry and told him 'that I would not let him separate us, that we had survived together through so much, and we would not lose each other now!' Soon after, they were transferred together to a hospital building in a former SS dormitory, where Halina was thrilled to see beds with clean sheets, blankets, and pillows.[9]

Halina Neujahr and her sister Sara, 26, were split up when they were transferred to newly established hospital barracks a few weeks after liberation, though Halina was not fully aware of what was happening. 'I felt like I was dead,' she later recalled. 'I was blue up to about the knees. I didn't feel any circulation in my body.' From her hospital bed, she could hear the cannon fire that signalled the end of war in Europe on 8 May, though the significance of it was lost on her at the time. As soon as she was able to walk, Halina located Sara, who was suffering from a high fever. The sisters had managed to stay together since their arrest in the Warsaw Ghetto in 1943. But their mother, Gitla-Ejdla Neujahr, was dead at the age of 47, almost certainly gassed at Majdanek soon after their own arrival there. According to Halina, while in Majdanek, she and Sara met a woman who had sheltered with them during the Ghetto Uprising.

> She told us that a few days after our capture, the hideout was discovered and everybody deported to Majdanek. Our mother was with the others. She was very depressed, mourning her lost children. I suppose that my mother didn't pass the selection at the intake to Majdanek. Her hair was already half grey. Her body was no longer that of a young girl. It seems that the selecting officers sent her directly to the gas chamber in the camp of Majdanek.[10]

In June, Halina and Sara Neujahr were asked if they would like to be repatriated to Poland or sent to Sweden, where they would receive medical treatment. Having lost their father years before and aware that Warsaw had been almost completely destroyed, the two young women knew there was nothing left for them in Poland. Besides, Sara had once heard wonderful things about Sweden from a teacher who studied there before the war. From Bergen-Belsen, they were loaded on stretchers onto a Red Cross train, which took them to Lübeck, Germany, where they would be received by the Swedish officials and taken to Sweden by boat. A highlight of this journey, Halina recalled, was when they were given blueberries and cigarettes. For some reason, Sara – who loved blueberries – got the blueberries while Halina – a non-smoker – got the cigarettes. After Sara finished her blueberries, someone offered Halina blueberries for her cigarettes – a valuable commodity. 'I was so selfish. I didn't want to sell the cigarettes to get more blueberries to give to my sister. And I feel guilty about that for the rest of [my] life,' Halina said later.[11]

The sisters arrived at the Swedish Transit Hospital in Lübeck on 27 June 1945. Like all former prisoners, they had to go through a series of medical checks and disinfection of their bodies and their clothing before they could board the ship for transport to Sweden. The disinfection process started with taking the clothing, which was disinfected separately, and placing them naked into a hot sauna. Next, they were given a cold shower, dried, and then sprayed from head to toe with DDT powder – a pesticide now considered toxic. Finally, parts of the body covered with hair were washed with insecticide. Sara went through this process first. 'She went ahead, and then I came and suddenly she was lying there,' Halina explained. 'And she said, "I don't want to die. I don't want to die"... And I kissed her on the mouth... and then she was dead.' Sara was buried in a Lübeck cemetery the following day. The day after her sister's burial, Halina was carried on a stretcher aboard a ship bound for Norrköping, Sweden, where she arrived on 1 July 1945.[12]

The same day Halina Neujahr arrived in Sweden, Halina Czajkowska was told that she and her mother might be able to go there as well. Like Halina and Sara Neujahr, Halina and Emilia were aware that there was little to return to in Poland. 'We knew that Poland was now occupied and controlled by the Russians who had imposed a Stalinist-Communist government,' she wrote later. 'We wondered where and how Father was, and we did not want to find ourselves on opposite sides of the world.' But Sweden was a complete unknown to them. 'The possibility of living in Sweden had never occurred to us, and we were uncertain, even fearful, about heading to another country.' Ultimately, their physician helped alleviate their fears, and they decided to take a chance on Sweden.[13] After a three-day journey from Lübeck, they arrived in Gothenburg, Sweden on 9 July 1945.

* * *

Even as they arrived in Sweden in the summer of 1945, survival was not certain for former prisoners like Halina Czajkowska and Halina Neujahr. In July and August 1945, nearly 10,000 former concentration camp prisoners were transported from Bergen-Belsen to Sweden via Lübeck. They arrived by boat to five Swedish port cities – Norrköping, Gothenburg, Stockholm, Malmö, and Kalmar. Cemeteries in each of these cities hold the graves of dozens of former prisoners who died either on the voyage or soon after arrival.[14] Many were young women, often only teenagers. Halina Czajkowska called this period of her life purgatory, which she described as 'a state of suffering and cleansing, which, sooner or later, will change definitively to a better or a worse state'.[15]

Problems at Home

On 3 June 1945, almost a month after the factory in Lenzing was liberated, Ellen and her mother Dina began their journey back to the Netherlands. Their Amsterdam home and business were no longer theirs, long ago confiscated by the Nazis and given to someone else. Julius Loeb had died in Westerbork. Ellen's sister Trudy Shakno and her family were in Dallas, Texas. Nonetheless, they were happy to be making the journey. On the way, they were treated with great kindness by Red Cross officials, who ensured that all their needs were met and provided a few luxuries. 'All the food yards were at our disposal,' Ellen wrote to Trudy. 'We hadn't eaten so much food in six years.' Throughout the train journey, they were showered with food, chocolate, and cigarettes by locals, and similar items by the Red Cross.[16] Their arrival in Amsterdam was greeted with joy by family friends, who wrote to Trudy in Dallas, 'It was for us an unforgettable moment when Dina and Illa [Ellen] one nice evening suddenly appeared in our door. Even the day before they were positively identified as being dead.'[17]

As wonderful as all of this must have been, Ellen and Dina had to contend not only with the consequences of their ordeals during the war, but also with daily life in postwar Amsterdam. 'Money has no value; only the black market operates at ridiculous prices,' Dina wrote to the Shaknos in September 1945. 'You won't believe that anti-Semitism has increased. We, who are stateless Jews, feel it every day. Haven't we suffered enough?'[18] Dina's observation about increased antisemitism was accurate. In fact, antisemitism had been increasing in the Netherlands during the German occupation and reached a peak as surviving Jews returned after the end of the war. One reason for this hostility was a sense of collective guilt within the Dutch population for not doing enough to help the Jews. As Jews returned to the Netherlands, the non-Jewish population 'sought to justify themselves by criticising the Jews,' writes historian Evelien Gans. The worst rebukes came with an especially painful sting, as Gans explains:

> The superlatively aggressive way of treating Jewish people was expressed in the abusive remark, 'They have forgotten to gas you.' This new stereotype –

the Jews are there to be gassed – was inspired by the persecution of the Jews itself. It was the worst thing you could say to a Jew, and that is how it was felt.[19]

Unsurprisingly, Ellen and Dina were anxious to leave the Netherlands.

* * *

Ruth made her journey back to Prague soon after the liberation of Theresienstadt in early May 1945. She had lost track of her husband Alfred in October 1944, during the selection at Auschwitz-Birkenau, when he was beaten by Nazi guards for getting out of line to give her a piece of chocolate. Unknown to Ruth, this act saved him from an immediate death in the gas chamber. It was not until she arrived in Theresienstadt in the late spring of 1945 that she learned Alfred had survived and was receiving medical treatment in Bergen-Belsen, which had been converted to a displaced persons (DP) camp after liberation in April. Ruth tried to communicate with Alfred, but her letters never reached him, and it wasn't until he returned to Prague that he discovered through a former colleague that she was alive. Once reunited in Prague, the couple initially lived in the apartment of a non-Jewish cousin. Alfred was still recovering from typhus and was, according to Ruth, a 'wreck'.[20]

Czechoslovakia was also a wreck. With the fall of Nazi Germany, Czechoslovakia established itself once again as a republic – the Third Czechoslovak Republic – encompassing its prewar borders. Soon, however, President Edvard Beneš was pressured to cede part of its restored territory, Carpathian Ruthenia, to the Soviet Union. The new coalition government included a strong Communist party that increasingly brought Czechoslovakia under the influence of the Soviet Union. In the meantime, the government was cleansing the country of ethnic Germans, resulting in the expulsion of around 2.5 million people between 1945 and 1947.[21] The situation was *Lebensraum* in reverse; a forced removal of Germans – or at least those considered German by countries still reeling from German occupation and persecution – back to the now-reduced borders of Germany.

In addition to xenophobia and the growing influence of Communism, there was antisemitism here too. Most of the 35,000 Jews who had lived in

Prague before the German occupation perished during the war. Ruth and Alfred were among a mere 5,000 Jews who survived the Nazi camps and returned to the city after the end of the war.[22] The return of these individuals was not celebrated. 'By and large, Czech Gentile society reacted with little patience or encouragement toward survivors,' writes Czech historian Anna Hájková. 'Postwar Czech society was highly nationalized and almost hysterically anti-German, celebrating its political prisoners and cherishing the myth of resistance. Jews were a foreign body, and it was up to them to find a way to fit in.'[23]

Ruth and Alfred did their best under the circumstances. When they were offered jobs in a private psychiatric hospital in Prague, they jumped at the chance, even though neither specialised in psychiatry. The jobs came with a small apartment inside the hospital, and that meant that they could at least live independently. But they had no intention of settling down permanently in Czechoslovakia. According to Ruth, the first day the United States Embassy in Prague opened in on 29 May 1945, they were there to register for visas to emigrate.[24]

* * *

Whereas Soviet influence was increasing in Czechoslovakia in the immediate aftermath of the Second World War, it was already firmly established in Hungary. German occupation was replaced with Soviet occupation, and deportation of Jews to Nazi concentration and labour camps was replaced with the deportation of Jewish and non-Jewish civilians and soldiers to Soviet labour camps. The Swiss Embassy in Budapest reported in March 1945 that a large concentration camp had been established outside of Budapest 'with about 40,000 internees, who are being deported to an unknown destination in the east. It is common knowledge that these internees are given little food unless they sign a statement that they wish to join the Red Army or to work in Russia of their own free will.'[25] An estimated 700,000 Hungarians were sent to Soviet Gulag camps in 1944 and 1945. Around one-third of these never returned.[26] Among those deported was Eva's brother. Arrested on the street one day, he was sent to Siberia, and the family lost all contact with him until he returned four years later.[27]

Those who remained behind were subject to other forms of violence, including rape and looting. 'After the fall of Budapest, [Soviet general] Malinovsky granted his troops three days of "free looting" – which included forcing women to act as prostitutes and holding them captive for a fortnight if they were considered attractive – to celebrate their victory.'[28] These activities did not end after the three-day grace period. Homes, businesses, and infrastructures were destroyed, leaving many civilians without homes and with limited access to water and food. The population of Budapest dropped significantly during the first months after liberation as citizens were rounded up for labour or escaped the city.[29] Twenty-year-old Eva was among them. She and her friend Clara made a dangerous journey to the city of Szeged, Hungary, around 180 kilometres southeast of Budapest, to find their boyfriends, who had been arrested during a labour round-up but managed to escape and flee to Szeged.

Carrying all their belongings in backpacks, without much money or food, the two women made the journey through whatever mode was available to them at any given time. Sometimes they walked, other times they rode on the top of crowded train cars. Most dangerous of all was accepting rides from Soviet soldiers. During one ride in a Soviet truck, a soldier took a liking to Eva and began speaking to her in an intimate way. Eva deflected his advances by asking him to help her learn to count in Russian. This seemed to distract the soldier until their next stop, when Eva and Clara managed to escape to the house of some acquaintances, where they stayed for the night. Eva avoided almost certain rape, but the solider had his revenge the next morning when he grabbed her backpack and drove away with it. The women resumed their journey to Szeged on foot, frightened but unharmed.[30]

* * *

The end of the Second World War is usually depicted as a joyous time. And though it was in some respects, the initial moments of joy were followed by the reality of the destruction caused by the war and the seemingly impossible task of reconstruction and renewal. It was also a time of tremendous political upheaval, economic insecurity, and social strife. Countries formerly occupied by Nazi Germany, such as the Netherlands, were struggling to reconcile the actions of their politicians and citizens in relation to the Nazis and the genocide of the Jews. A chasm between East and West was growing with the Soviet Union on the one side and the other former Allies, led by the United States, on the other. Countries that were coming under the influence of the Soviet Union, like Czechoslovakia and Hungary, faced increasing isolation from the Western world. For Ellen, Ruth, and Eva, 'home' was not the place it had once been. Their purpose now was to make new lives for themselves. The questions were how and where to do that.

Seeking Purpose

Rita had known from a young age that she did not want to follow the traditional path for women of her class and time. Marriage and family were not for her. The problem was that she had no conception during her youth of what she could do with her life since she believed that, unlike her sisters, she lacked talent and inclination. The reality was that it was her education, not her ability, that was lacking. 'The biological sciences were not part of our high school's curriculum,' she recalled in her memoirs. 'To me, a bat was not different from a bird, both being flying creatures; nor had I any clear ideas about the difference between the various inhabitants of fresh and salt water, from crustaceans to fish or whales.' In 1930, without any knowledge of science or medicine, the death of her beloved governess prompted her to decide to study medicine. By her own account, this purpose became 'the very reason of my existence'.[31] She fulfilled her ambition in 1936, when she earned her medical degree with top honours. Just two years later, in 1938, the Italian racial laws prohibited her from practising medicine. After a brief attempt at a clandestine medical practice, she turned her attention solely to research until the liberation of Florence gave her another chance to work as a physician.

In early September 1944, Rita volunteered her services as a physician to the Allied forces and was assigned to work at an emergency field hospital treating war refugees. During the summer, tens of thousands of civilians fled to Florence from the surrounding countryside, where fighting between the Allies and Axis forces had been intense, and more were brought to the city for refuge and medical care that fall and winter. The experience changed the course of her professional career. Treating war refugees was painful and frustrating, leaving her with a feeling of impotence. 'The consequences of malnutrition and of the cold were particularly evident in babies and old people,' she explained. 'Many of the newborn children arrived in a state of extreme dehydration. At dawn I would take them to the pediatric clinic, only to be an impotent witness to the extinction of the lives of these small creatures, by then already on the threshold of death.'[32] Overcrowding in the makeshift hospital, which Rita characterised as little more than a shed, led to the rapid spread of typhoid fever among the refugees in late winter 1945. It was a turning point in 35-year-old Rita's life. Although she continued working as a physician through the spring of 1945, her experiences ultimately contributed to her decision not to practise medicine. Many years later, she reflected that she lacked 'the detachment that allows a doctor to face the suffering of a patient without creating an emotional involvement damaging to both parties'.[33]

With the end of the war in Europe, Rita returned to Turin in July 1945. Deeply affected by her experiences as a physician, she simultaneously lost her enthusiasm for her research.

> With the disappearance of the danger that had threatened the survival of many of us, so also disappeared the strength or blind daring that enables one to overcome life's most difficult moments. I fell into a depressed state similar to those so frequently manifested during convalescence from long and serious illnesses.

Rita's depressed state was compounded by a sense of inadequacy about herself and her scientific training. Her solution was to enrol in biology courses to address the gaps in her education.[34]

* * *

'I'm not happy about myself,' Ellen wrote to Trudy in late July 1945. 'I'm now almost 26 years old and still do not have a finished education. It upsets me to no end.'[35] Ellen's formal education was interrupted by the Nazi occupation of the Netherlands and her deportation to Westerbork in late 1942, where her medical career effectively began. She spent nearly three years working as a nurse and sometimes even in the capacity of a doctor. But when she walked out of Lenzing in May 1945, she had no formal education or qualifications in either capacity. Her experience in the camps was unlikely to be considered as valid by any medical school or institution. Ellen's immediate ambition was to get the training needed to qualify as a nurse. 'I'm planning to work as soon as possible as a nurse,' she wrote to Trudy earlier in July. 'I won't make much money, but I need a diploma.'[36] Fortunately, nurses were desperately needed in the Netherlands to fill the postwar shortage that occurred just as demand increased.[37] She could begin studying for a nursing diploma in the Netherlands, but completing the programme would take at least two years, and she hoped to be in Dallas before then. Given the conditions in the Netherlands, Ellen and Dina were anxious to leave. Their preference was to go to Dallas, where Trudy lived with her family. Ellen even turned down an offer to emigrate to England, where there was the possibility of attending nursing school.[38] At the same time, she had greater ambitions than being a nurse and working in the Netherlands. 'You might think I'm deranged, but deep in my heart, I would like to study medicine,' she wrote to Trudy. 'I know you think I've lost my senses after what we went through to plan for a future as a physician, but we're free now, and this is my great wish.'[39]

Trudy and her husband had become US citizens, which enabled Trudy to apply for her mother and sister to come to the US on a preferred immigration quota for relatives of citizens as soon as she knew they had survived.[40] However, emigration was difficult even with relatives in the US, mainly due to issues of transportation. 'Here is nothing [sic] known about transportation to the U.S., but a few go with freighters,' Dina wrote to Trudy in September 1945. Compounding matters was the fact that Ellen and Dina were still recovering from malnutrition and illness from their time in the camps. They were also occupied with recovering whatever personal belongings they could and applying for restitution of the life that was stolen from them by the Nazis. Despite the challenges and uncertainty, according to Trudy Shakno, Ellen somehow managed to enrol in medical school in the Netherlands.[41]

* * *

'Nurses attend to starving children in a hospital in the Warsaw ghetto'. Circa 1942, Warsaw, Poland. United States Holocaust Memorial Museum, courtesy of Maladie de famine: recherches cliniques sur la famine… dans le ghetto de Varsovie – Apfelbaum, Emil (ed.) – American Joint Distribution Committee.

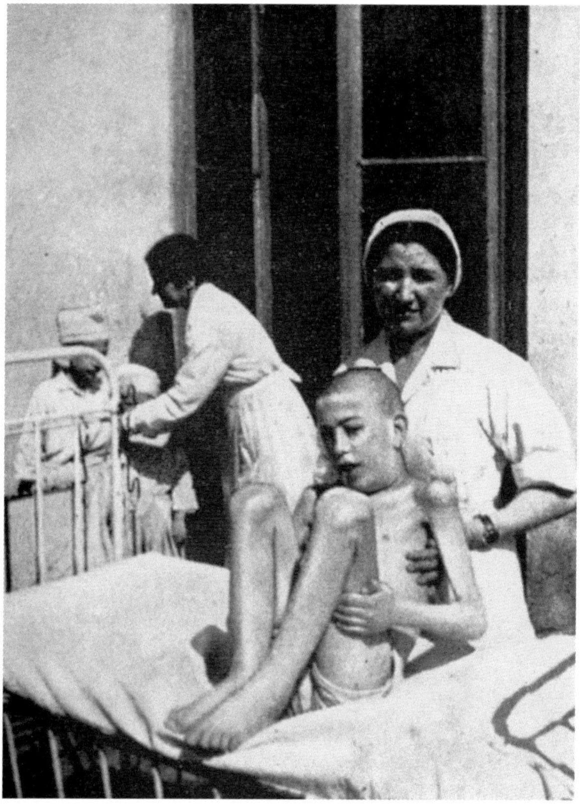

Rita Levi-Montalcini in 1948. Unknown location and photographer. From the Viktor Hamburger Collection of the Marine Biological Laboratory of the Woods Hole Institute, https://hdl.handle.net/1912/19757. CC-BY 4.0 International license.

Dr. Ruth Pick with Dr Louis N. Katz in the laboratory at Michael Reese Hospital, Chicago. The imag accompanied journalist Lillian Pompian's 24 May 1959 *Chicago Tribune* article, 'Your Heart is in Thei Work'. Andrew Pavlin/Chicago Tribune/TCA. Reprinted with permission.

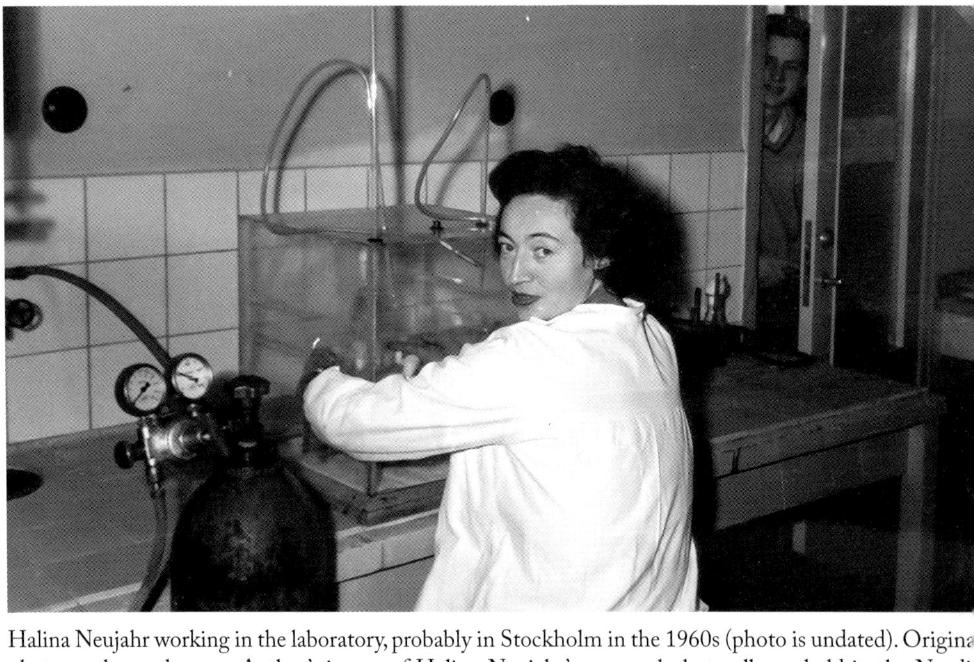

Halina Neujahr working in the laboratory, probably in Stockholm in the 1960s (photo is undated). Origina photographer unknown. Author's image of Halina Neujahr's personal photo album, held in the Nordi Museum (*Nordiska Museet*), Stockholm, Sweden.

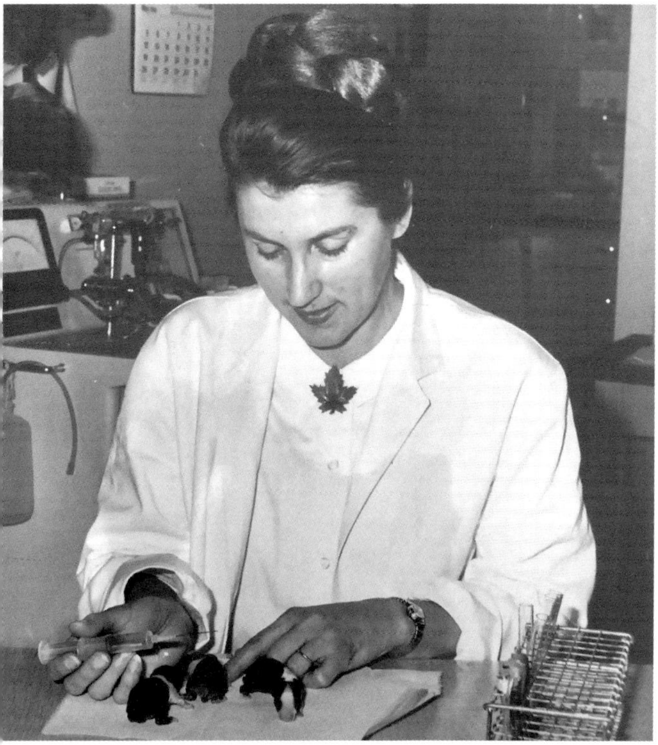

Halina Czajkowska in the laboratory at Karolinska Institute, Stockholm, Sweden, circa 1950. She is working with some of the mice brought back from the United States by Georg Klein. Coincidentally, she is wearing a maple leaf brooch, foreshadowing her future in Canada. Photographer unknown. Halina Czajkowska Robinson Collection, courtesy Jacalyn Duffin.

The village of Lidice in Bohemia, 16 kilometres from Prague, which was destroyed by the Germans in retaliation for the assassination of Reinhard Heydrich, acting governor of the Protectorate of Bohemia and Moravia. Image from June 1942. United States Holocaust Memorial Museum, courtesy of Library of Congress (Public Domain).

Twenty-year-old Rita Levi, as she was then known, as a first-year medical student at the University of Turin in 1930. In the collection of the Historical Archive of the University of Turin. Reprinted with permission.

Westerbork Transit Camp in The Netherlands during the Second World War, circa 1944. Unknown photographer. Public domain image (Wikimedia Commons).

'Warsaw Ghetto Uprising (Poland) – Jewish women and children forcibly removed from a bunker by Schutzstaffel (SS) units for deportation either to Majdanek or Treblinka extermination camps (1943)'. Photo from Jürgen Stroop Report to Heinrich Himmler from May 1943. Public domain image (Wikimedia Commons).

Selection at Auschwitz-Birkenau, May 1944. 'Jews from Subcarpathian Rus await selection on the ramp at Auschwitz-Birkenau', United States Holocaust Memorial Museum, courtesy of Yad Vashem (Public Domain).

The arrest of Jews in Budapest, Hungary, October 1944. From the collection of the German Federal Archives. Bundesarchiv, Bild 101I-680-8285A-26 / Faupel / CC-BY-SA 3.0.

'Budapest, Hungary, 1945. The burnt-out remains of Buda Castle overlooking the destroyed Chain Bridge' Photographer: Kurutz Márton. Wikimedia Commons, CC BY-SA 3.0.

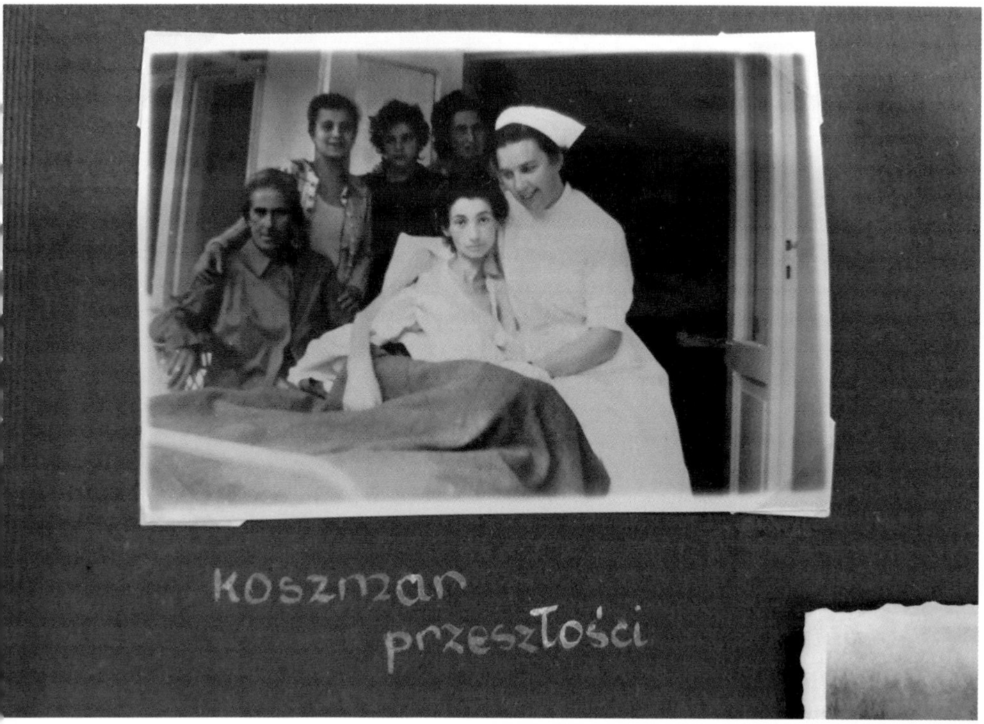

A page from Halina Neujahr's personal photo album. She is the emaciated figure in the hospital bed. The photo was taken soon after her arrival at the emergency field hospital in Norrköping, Sweden on 1 July 1945. She captioned the photograph (originally in Polish), 'A Nightmare of the Past'. Original photographer unknown. Author's image of Halina Neujahr's album, held in the Nordic Museum (Nordiska Museet), Stockholm, Sweden. My thanks to Roman Wroblewski for helping to confirm the location and date.

A page from Halina Neujahr's personal photo album, showing her looking much healthier than the previous photo. She captioned the photograph (originally in Swedish), 'Toward a new life. Summer 1945'. At the time she would have been preparing to leave the emergency field hospital in Norrköping, Sweden. Original photographer unknown. Author's image of Halina Neujahr's album, held in the Nordic Museum (Nordiska Museet), Stockholm, Sweden.

The refugee camp in Öreryd, Sweden, circa 1945. Photographer unknown. Courtesy of Öreryds Hembygdsförening (Public Domain).

Photograph page from the passport issued to Ilse Loeb, as she was then known, by the *Commissaris van de Koning* in Haarlem, Netherlands, dated 14 August 1946. From the Dr. Ellen Loeb-Katz Collection Dallas Holocaust and Human Rights Museum. 2002.34.61. Reprinted with permission.

Halina Czajkowska in Sweden, circa 1948. Photographer unknown. Halina Czajkowska Robinson Collection, courtesy Jacalyn Duffin.

A page from Halina Neujahr's personal photo album, featuring a photo of a formal gathering of KTH students (including Halina, bottom row center) called 'Nollegasque' in Swedish. This event marked the end of the first year of study for a new class of students – referred to as 'nollor' or, in English, 'zeros' (hence the costumes). This is one of several photos in her album that capture this important milestone in Halina's education, circa 1950. The printed caption reads in translation from Swedish, 'Nollegasque around 1950. Photo in private collection [possibly Halina's]. The lonely girl is Halina Neujahr'. Original photographer unknown. Author's image of Halina Neujahr's album, held in the Nordic Museum (Nordiska Museet), Stockholm, Sweden.

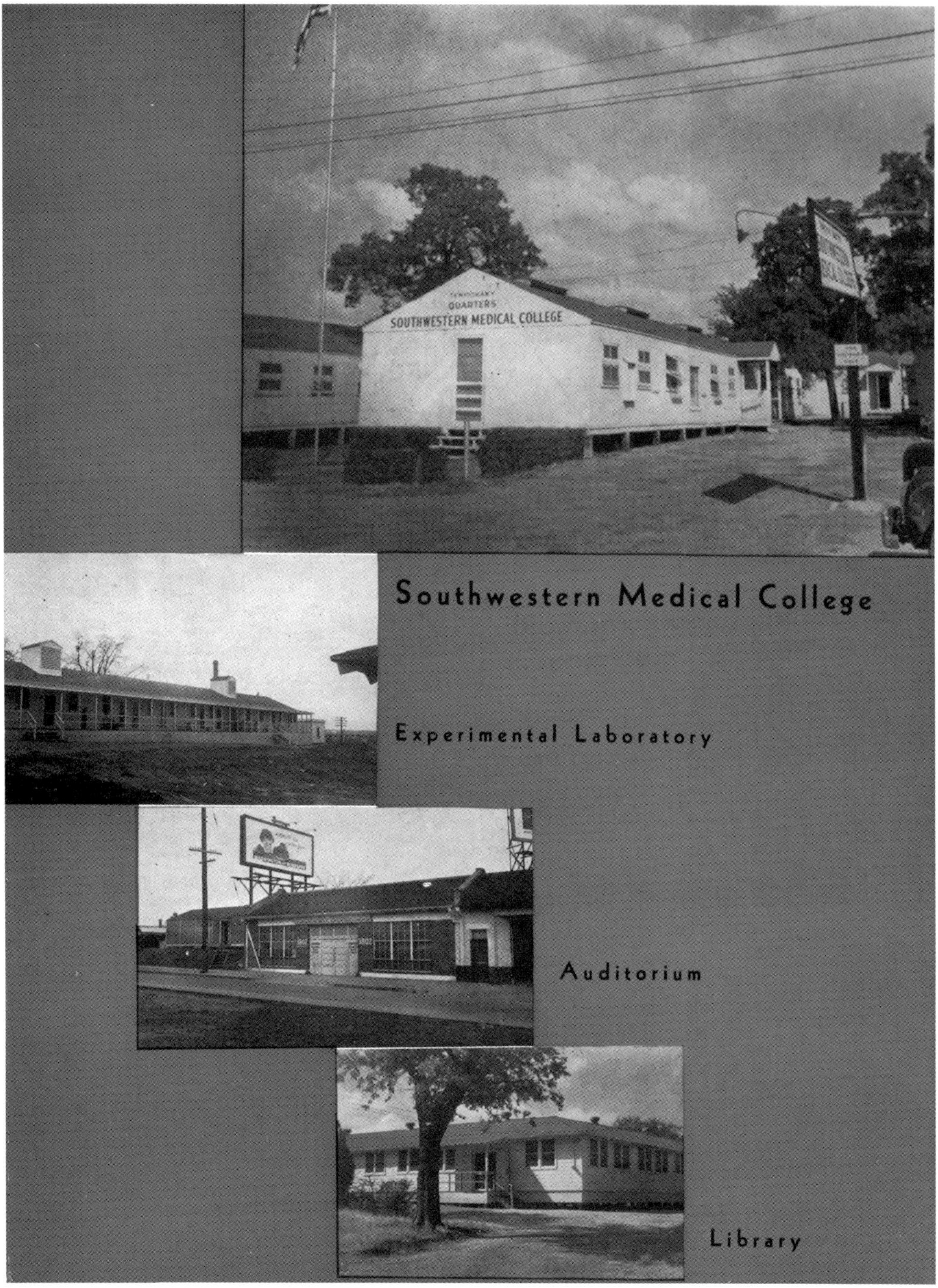

A page from the Southwestern Medical College yearbook for 1949, the year Ellen Loeb started her medical studies, featuring several buildings on campus. Caduceus. 1949. Yearbook Collection. 1927-Present. MSS0057. University of Texas Southwestern Medical Center Special Collections Library and University Archives. Dallas, Texas. Reprinted with permission.

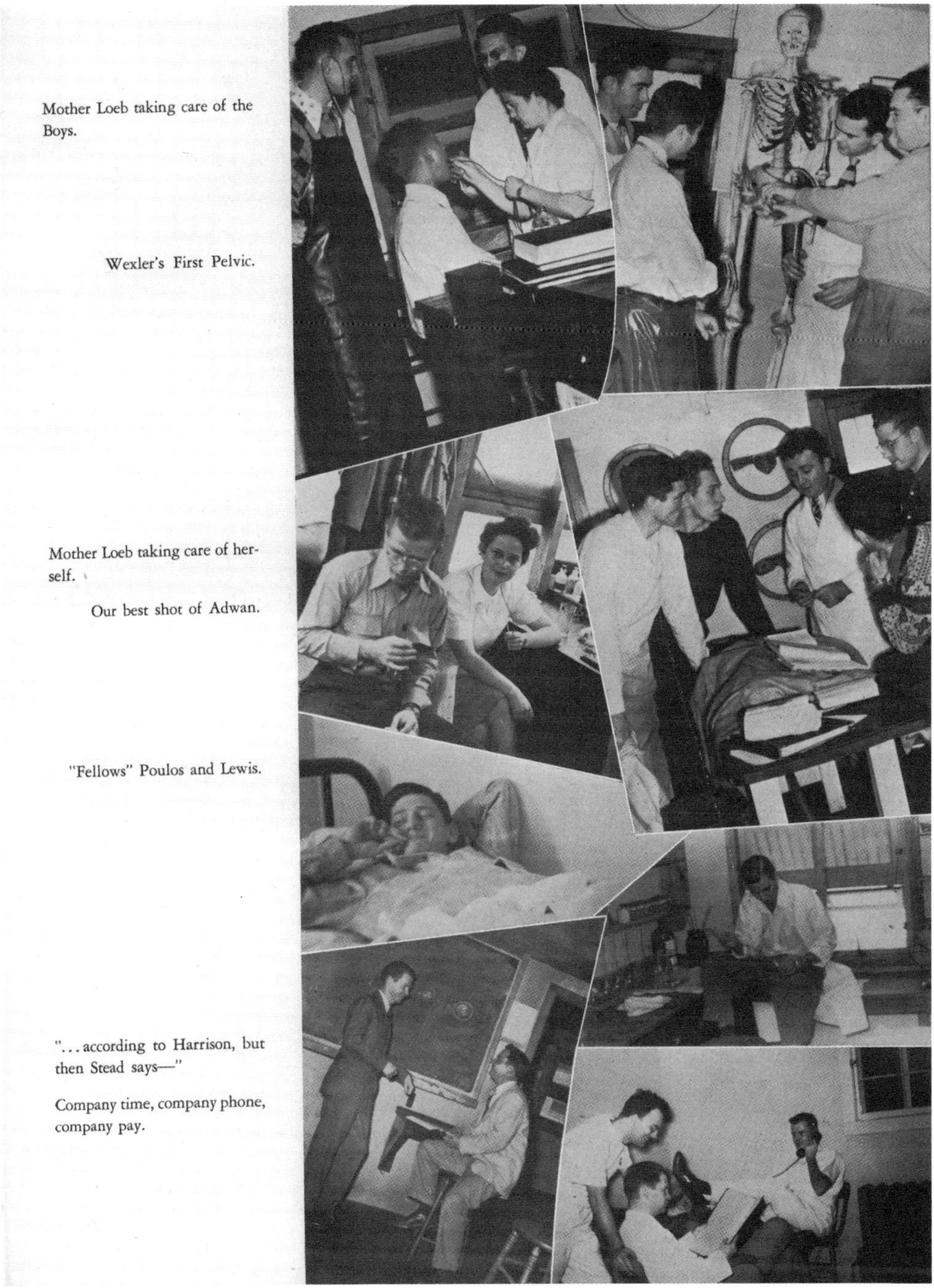

A page from the Southwestern Medical College yearbook for 1949, featuring Ellen Loeb in two photos; one captioned 'Mother Loeb taking care of the Boys' and another captioned 'Mother Loeb taking care of herself'. Caduceus. 1949. Yearbook Collection. 1927–Present. MSS0057. University of Texas Southwestern Medical Center Special Collections Library and University Archives. Dallas, Texas. Reprinted with permission.

Southwestern Medical School, class of 1952 group photo. 1949. Ellen is seen standing in the front row. From the University of Texas Southwestern Collection. 1943–Present. MSS0021. University of Texas Southwestern Medical Center Special Collections Library and University Archives. Dallas, Texas. Reprinted with permission.

Viktor Hamburger and Rita Levi-Montalcini on Mont Blanc, 1967. Photographer unknown. VC089-i089021, Bernard Becker Medical Library Archives, Washington University in St Louis.

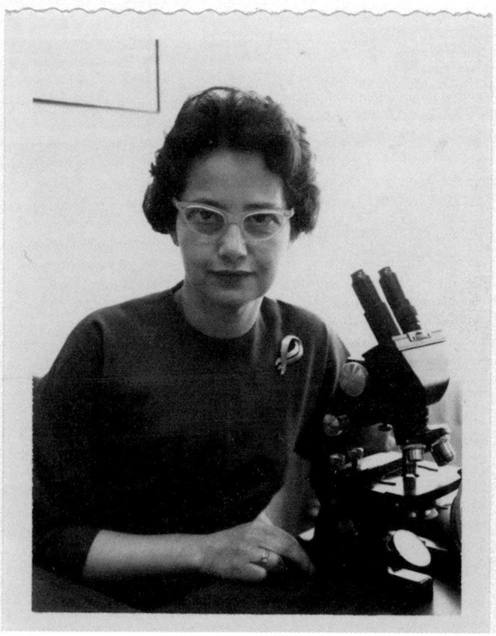

Dr Ellen Loeb circa 1964. Unknown photographer. From the Dr Ellen Loeb-Katz Collection, Dallas Holocaust and Human Rights Museum. 2002.34.76. Reprinted with permission.

USC Shoah Foundation screen grab of Ruth Pick giving testimony in Chicago, 21 February 1995. Reprinted with permission.

Portrait of Rita Levi-Montalcini in her laboratory, sitting behind a microscope, circa 1963. Photographed by Herb Weitman. VC089-i089010, Bernard Becker Medical Library Archives, Washington University in St Louis. Reprinted with permission.

Viktor Hamburger and Levi-Montalcini at the Fidia Awards, 1987. Unknown photographer. From the Viktor Hamburger Collection of the Marine Biological Laboratory of the Woods Hole Institute, https://hdl.handle.net/1912/19764. CC BY 4.0 International.

Wedding portrait of Halina Czajkowska Robinson and James Russell 'Rus' Robinson, 1 August 1953. Photographer: Victor Aziz, London (Ontario). Halina Czajkowska Robinson Collection, courtesy Jacalyn Duffin.

Eva and Georg Klein, Karolinska Institute, Stockholm, 11 April 1979. Unknown photographer. This image was released by the National Cancer Institute, part of the National Institutes of Health, with the ID 8175, and is in the public domain.

Halina Neujahr (standing) towards the end of her life, hospitalised and suffering from dementia. Her friend and former colleague from KTH, Roman Wroblewski, was visiting her when he took this photo. Of this photo, he writes: 'One day, in the patient cafeteria, I met a handsome, tall man with the same diagnosis as Halina, dementia. He asked me where Halina was from. I said she was born in Warsaw. Suddenly the man's face beamed with happiness. He went to the piano in the corner and began to play the Song of Warsaw. A song that was the most played melody in all of Europe after the war. Halina's tense face also changed. Completely relaxed, she went to the playing man and put her hand on his shoulder. It was an incredibly powerful experience. I left them alone.' My thanks to Roman Wroblewski for allowing me to reprint the photograph here.

Halina Czajkowska Robinson in 2019, photographed by Jacalyn Duffin. Reprinted with permission.

Eva and Georg Klein in 2008 at the Nobel Forum, Karolinska Institutet, photographed by Stefan Zimmerman/ Karolinska Institutet. Reprinted with permission.

Eva, then aged 20, and her friend Clara had travelled from Budapest to Szeged to join their boyfriends. But when Eva told her story to the USC Shoah Foundation in 1997, she made no mention of a reunion with her boyfriend. Instead, she explained that it was in Szeged that she started medical school.[42] Her future purpose was still uncertain, however. As she had before the war, she vacillated between a career in medicine or as an actress. What she wanted, she later reflected, was far simpler than either of these two possibilities. 'We emerged from the chaos after the war – we felt reborn – I did not wish for anything else but a well ordered life,' she later wrote.[43] In Szeged, she dropped out of medical school to join a theatre group. But even this did not last long, and soon she returned to Budapest with her mind made up that life as an actress was not what she wanted.[44] Before the Second World War, the possibility of attending university had been restricted for Eva and other Jews in Hungary under the *Numerus Clausus* law. As difficult as life in early postwar Hungary was, she was now at least able to exercise her options for study.

In Budapest, Eva again enrolled in medical school. It was at this time, she later said, that her 'proper life as a medical student starts'.[45] The personal uncertainties of her teenage years behind her, she had seemingly found her purpose just as life in Hungary became more uncertain. Ideologically, Hungary was the target of a tug-of-war between the United States and the Soviet Union, with democracy on one side and Communism on the other. The Cold War was imminent, and all signs seemed to indicate that Hungary would be on the Soviet side of the Iron Curtain.[46]

* * *

During the Christmas holidays of 1945, a Polish professor addressed Halina Czajkowska and other liberated prisoners in a Swedish convalescent home about their futures. 'I asked him about the prospects for people of my age, who had worked so hard during the war to advance their education, only to lose so much time as prisoners, and who would now like to go to school again,' Halina explained in her memoirs. 'He asked many questions in return, made notes, and left with a vague promise that I might hear from him in the near future.'[47]

Halina had spent much of the war afflicted with various illnesses, threatening both her education and her life but also instilling in her a certain defiance against both threats. Before her deportation to Auschwitz-Birkenau in August 1944, she had spent much of the year ill, including suffering a burst appendix in June, which was removed on her 17th birthday, 23 June 1944. In her memoirs, Halina attributes her ill health that year in part to her preparations to complete her junior matriculation, which she felt caused her to have low resistance to infection.[48] After arriving in Sweden in July 1945, she was hospitalised for acute pericarditis – inflammation of the pericardium, the lining that surrounds the heart. Her treatment and recovery lasted more than three months, which she spent in a hospital in Gothenburg, Sweden. There, she and her mother learned that Halina's father had survived the war.[49]

With her health and family in order, Halina was transferred to a convalescent home, where her thoughts naturally returned to her education. Her persistence paid off when, soon after Christmas, she learned that she was being sent to a school established for young Polish refugees deep in the forests of southern Sweden. She arrived in the small church village of Öreryd on 9 January 1946. A refugee camp was established in the village at the beginning of the Second World War. During the war, it housed Norwegians fleeing the German occupation of Norway. Within weeks after the end of the war, these individuals had returned home and the site became a place of refuge for some of the thousands of Polish former prisoners of Nazi concentration camps who came to Sweden in the spring and summer of 1945.[50] Later that year, a school was established for young Poles like Halina Czajkowska. And Halina Neujahr.

* * *

'I came to Sweden, severely ill, weighing slightly more than 30 kg and alone. My sister Sara, to whom I owe my survival through the camps, was left behind in a mass grave, in Lübeck. The newly won freedom, all the beauty of life now available, suddenly lost its value.'[51] Like Halina Czajkowska, Halina Neujahr suffered in both body and mind during the Second World War and the Holocaust. On 1 July 1945, 20-year-old Halina N. was carried off the boat on a stretcher at the harbour in Norrköping, Sweden, and immediately transferred to an emergency field hospital set up in a local school that was closed for the summer holidays. In addition to being severely malnourished, she had lost her hair due to typhus and was suffering from pleurisy – inflammation of the tissue separating the lungs from the chest cavity. With no family or friends, Halina found comfort in the kindness of strangers. Cut off from all contact with the outside world due to strict quarantine rules, Halina and the other patients received gifts of chocolate and flowers from the local population and churches. Despite the small comforts these kindnesses provided, Halina was suffering from the traumas she experienced. While in the hospital in Norrköping, she wrote about her experiences during the Holocaust in an evocative short story titled in Polish *Oczy świata* (in English, 'The world's eyes') that reveals the continued angst she felt in the aftermath.[52]

In September 1945, the school in Norrköping was needed for the autumn term, and so Halina was transferred to a sanatorium in the village of Rosöga. There, she began to regain some of the 'beauty of life' that she had lost in Poland and Germany through what she later described as hard work, study, and research.[53] In January 1946, she volunteered to record in writing the experiences of three other Jewish and non-Jewish survivors of Nazi persecution receiving medical treatment in the sanatorium for an initiative dedicated to creating a record of Polish suffering during the Second World War and the Holocaust.[54] While there, she heard about the Polish school in Öreryd and was determined to go. Her doctors were hesitant, however, and initially refused. Finally, in June 1946, the doctors gave her the green light, and Halina and another young woman from the sanatorium boarded a train to Öreryd.[55]

* * *

Having just survived genocide, Ruth was now threatened with ethnic cleansing in Czechoslovakia. Because Carlsbad in Bohemia was part of the Austrian Empire when Ruth was born in 1913, her native language was German, even after the First World War, when Carlsbad became part of the newly formed state of Czechoslovakia as Karlovy Vary. The newly formed Czechoslovak Republic recognised national minorities, with nationality determined by the mother tongue of the individual. An exception was made for Jews, who could indicate 'Jewish' as their nationality even if Hebrew was not their mother tongue. In the Czechoslovak Census of 1930 – the last before the war – Ruth's husband Alfred gave his nationality as Jewish. But Ruth, who was a teenager at the time, took her father's advice and gave her nationality as German rather than Jewish on the basis that she spoke German, not Hebrew.[56] In early postwar Czechoslovakia, nationality was based on the 1930 census, with the consequence that Ruth was considered an ethnic German subject to expulsion. Fortunately, Ruth and Alfred were in a better position than many, and Alfred pressed his advantage. 'With lots of bribing, he got me all the papers necessary to stay, so I was not deported,' Ruth explained with a sly smile.[57]

The bribes were not made with money. 'Money did not mean a thing, unless you had [American] dollars,' according to Ruth. Or American cigarettes. The Picks were lucky that Ruth had an aunt in the United States who sent them things, including American cigarettes. Although she was a smoker at the time, Ruth did not smoke the valuable American cigarettes. 'They were high value money,' she recalled. 'Many of the papers we had to get… all was paid with cigarettes.'[58] Despite their tobacco wealth, life remained a challenge for the couple, not least professionally. Both disliked their work at the psychiatric hospital. But while Alfred, who specialised in cardiology, was able to find more favourable employment, Ruth could not. Having completed her medical studies just before the war, she did not have time to study for a specialisation. With fewer credentials, she had to take what work she could find, which meant staying at the psychiatric hospital. As much as she disliked the work, her employment continued to provide the couple with their own apartment and regular meals that would have been more difficult to obtain otherwise.

Life became significantly more difficult in February 1948, when the Communist Party of Czechoslovakia seized control of the government after steadily gaining power during the early postwar years. The couple lived in constant fear of arrest under absurd pretexts. Meanwhile, their efforts to emigrate proceeded frustratingly slowly. When the passports necessary to make the journey to the US finally arrived, they were taken away, returned, and then taken again.

Plagued with uncertainties and doubts about themselves and their possibilities – not to mention their precarious health and political status – each of the six women nonetheless were intent on improving their circumstances. There was no straight path ahead for any of them. Each was already on her way down a long and twisted road with no guarantee of success.

A Long Road

For survivors of Nazi persecution, 'home' was a complicated matter after the Second World War and the Holocaust. The physical building you once called home quite possibly no longer existed. If it did, someone else probably lived in it. Quite likely, many if not most of the people you once knew and loved were dead. As for your national home, it was almost certainly a different place than it had been before the war. Shifting borders, foreign occupation, overzealous nationalism, and divisive ideologies were like landmines buried in an unfamiliar and ruined landscape. For years after the end of the war, millions of people were 'displaced', stateless, homeless. Quite often, you had few if any choices about where you went after liberation. Your choice might simply be 'home' or somewhere about which you knew very little if anything at all. Both destinations could be nothing more than a temporary stop – maybe a respite, maybe a bad diversion – on the road to another more permanent destination. Sometimes no choice was required, you simply stayed where you had always been. In those cases, you might be like the proverbial frog slowly boiling alive as the water temperature gradually increases. Whatever the case may have been, the road to 'home' – wherever that might be – was often long and arduous.

* * *

Halina Czajkowska and Halina Neujahr both refused to return to Poland and were sent instead to Sweden, where they were expected to remain only temporarily, until they recovered their physical health enough to return to their native country or migrate elsewhere. Fortunately, by the time both women had sufficiently regained their physical health, Swedish policies had loosened to an extent, and they were not forced to leave. Even better, they were both given an invaluable opportunity to make up for lost time by returning to their studies at the Polish Folk School in Öreryd, Sweden, which had been established as part of an agreement between Swedish and Polish authorities. Between autumn 1945 and autumn 1946, nearly 200 Jewish and non-Jewish Polish survivors between the ages of 9 and 25 attended the school, which was composed of levels ranging from primary school to upper secondary school, following the Polish standard used at the time. The teachers were all Polish survivors, both Jewish and non-Jewish, who gave instruction to the students in the full range of subjects. Male and female students and teachers mingled freely although they lived in separate barracks for women and men.

Halina Czajkowska was the first to arrive, in January 1946, when she joined the highest level of the school, the 'lyceum', which was equivalent to advanced secondary education in Poland. Although she was three years younger than Halina Neujahr, she completed her studies at the school in September 1946, not long after 22-year-old Halina N. started hers in the summer of that year. Since Halina N. had missed three years of school during the war, she started at the 'gymnasium' level – the equivalent of compulsory secondary education in Poland.[59]

'Instruction started at 8 a.m. and continued with only a short break for lunch until 6 p.m. when dinner was served. Most days, we had another two hours of classes from 7 p.m. to 9 p.m. and the voluminous homework could occupy us until 2 or 3 a.m.,' Halina C. wrote in her memoirs. The result of this intense work was that she and the other lyceum students completed two years of study in one year and successfully passed their exams without the benefit of textbooks.[60] Under the same circumstances, Halina N. completed three years of lost education in one year. She was particularly inspired by mathematics and physics teacher Magister Piotr Eisenstadt (1907–1966), also a Jewish survivor. 'He was a genius,' Halina N. said, recalling that he taught the students everything by memory, since they did not have any books in Polish and, of course, had no knowledge of Swedish.[61] But the students were not driven to complete their studies as quickly as possible only by their passion for education. They also felt considerable pressure due to uncertainty about their migration status and future prospects. Halina C. gives some sense of their fears in her memoir:

> First, we were afraid that if we did not show our utmost effort, the Swedish government might cut off funding and other supports before we had completed our studies. Secondly, we felt that younger groups might benefit from our success and departure, giving them space and opportunity to continue their work to completion. Also, we wanted to set a good example.[62]

Despite the hard work, the students also managed to have fun. 'We made up for the intensive work on Saturday evenings with dance parties, featuring lively waltzes, polkas, and our favorite foxtrots,' wrote Halina C. 'Everyone danced with everyone else, as we worked out our frustrations and tensions from the hard concentration of the previous week.'[63] The two women almost certainly crossed paths at the school during the summer of 1946. Photos in Halina N.'s personal photo album document her time in Öreryd, making the most of abundant forests and lakes in the area.[64] On 14 September 1946, a graduation ceremony was held for the twelve lyceum students, including Halina C., who recalls a 'sumptuous banquet'.[65] Although it was certainly a time of joy, the occasion was also punctuated by continued uncertainty. The camp and school in Öreryd were closing, and Halina N. was transferred to another Polish school near Stockholm to finish her studies there. Halina C. was also still dependent on the generosity of the Swedish government and so was sent to the north of Sweden temporarily while the next step in her journey was decided.

* * *

In the immediate aftermath of the Second World War, immigration to the United States was no easier than it had been during the Second World War. The US, like much of the 'Western' world during the war – including the United Kingdom, Canada, and Australia – had limited immigration despite the desperate need of so many Jews and others fleeing Nazi persecution. As historian David S. Wyman wrote of Jews who sought but failed to find refuge in the US during the war in his seminal 1984 book, *The Abandonment of the Jews: America and the Holocaust, 1941–1945*, 'It was their particular misfortune not only to be foreigners but also to be Jews.'[66] Even as news of the liberation of Nazi concentration camps spread in the US and around the world in mid-1945, leading to widespread horror at the fate of millions of victims and the terrible state of the survivors, US immigration policy remained restricted according to a quota established in 1924 that reflected America's isolationist policies following the First World War. That quota not only limited the number of immigrants but also favoured immigrants from Western European countries like Britain, Germany, and Scandinavia. In late 1945, US President Truman ordered that the existing quota should be designated for displaced persons, but it wasn't until the Displaced Persons Act of 1948 that the limits themselves were addressed to enable the large number of stateless and other individuals who wished to settle in the US instead of returning to their former homelands.[67]

Ellen and her mother Dina were fortunate to have had the preferential treatment given to relatives of US citizens that would enable them to circumvent the restrictive quotas. Still, challenges remained – mainly related to bureaucracy and transportation. Trudy and her husband Rudy were not wealthy but were required to demonstrate that they would be able to financially support Ellen and Dina. In this, they were helped by their father's first cousin, Carl M. Loeb, who had come to the US in the late nineteenth century and since established himself as a successful American businessman. In his petition to the US authorities for Ellen and Dina's visa application, he was required to enumerate his assets. He indicated 'in excess of $1,000,000' and explained, 'The financial standing of my firm as well as my own is so well known, nationally and internationally, that I am not submitting corroborative evidence.'[68] As much as this must have helped expedite their visa application, Ellen and Dina still had to secure transportation to the US at a time when ships were the primary mode of transoceanic travel and US soldiers were being returned *en masse* from Europe. Finally, with their documentation in order, Ellen and Dina boarded the US ship the SS *Tillie Lykes*, which was transporting cotton, in Rotterdam on 1 October 1946. They arrived in New Orleans, Louisiana on 20 October 1946, ready to make the journey to Dallas, Texas to be reunited with Trudy and begin new lives.

* * *

In 1940, Rita had been inspired by Viktor Hamburger's 1934 journal article about embryonic nerve development. Her doubts about his conclusions led her to conduct her own, clandestine research to test Hamburger's theory. Two years later, according to Rita, her efforts 'yielded unhoped-for-successes'. Working with Giuseppi Levi, she came to a different conclusion that justified her scepticism about Hamburger's 'organizers'. She wrote in her memoirs that their findings 'suggested a different explanation from the one advanced by Hamburger to explain the almost complete disappearance of the motor cells in the spinal ganglia that innervated the limbs in embryos subjected to such destructive treatment'. The difference, Rita concluded, was the absence of a trophic factor rather than an inductive factor, as Hamburger had hypothesised.[69] In other words, Rita had discovered what controlled the fate of certain cells, including nerve cells. Because of the war and the fact that Rita and Levi were Jewish, these important findings were not published until several years later, and it took even longer for the impact of the findings to find Rita. The critical moment came in summer 1946, when Viktor Hamburger wrote a letter to Levi, having read about the findings in a Belgian journal. As Rita later explained, Hamburger was 'curious about my conclusions regarding the mechanisms governing the effects of peripheral tissues on the development of the nervous fibers that innervate them – conclusions completely different from his own'. Hamburger wanted Rita to come to St Louis for a semester to further study her findings with him.[70]

As satisfying as this was, Rita was determined to continue her biological studies, which she had begun in a quest to address her perceived inadequacies. She accepted the opportunity but delayed her departure for one year while she completed her studies. In September 1947, she sailed to the US from Genoa. Her plan was to stay in St Louis for no more than nine months.[71]

* * *

Around the time Eva made her journey from Budapest to Szeged with her friend Clara in early 1945, another would-be medical student, Georg Klein, made the same journey. In contrast to Eva, whose medical studies were interrupted by her experiment with theatre, Georg's medical studies continued unbroken. Eva and Georg met briefly in Szeged. As Georg later recalled:

> I had seen her before at the university, but my obsessive preoccupation with work prevented me from giving her or any other girl much attention. Still, I could remember very well how I met her the first time. On the second day of my medical studies in Szeged, I was standing in the Dean's office, to get my papers. She entered, dressed in a skiing outfit, having arrived in the city after a long and adventurous trip from Budapest, like my own. She asked me how to get papers. I saw that she was very beautiful. Her direct way of talking to a strange boy – very unusual for a girl in Hungary at the time – struck me as original and sympathetic.[72]

For her part, Eva thought Georg was very serious and studious. The two attended the same lectures for a short time until Eva dropped her medical studies to join the theatre group. When she returned to medical school in Budapest, where Georg had also enrolled as soon as the university reopened, they crossed paths once again through mutual friends. Then, in early summer 1947, Georg was invited to spend a week at a bombed-out summer house on Lake Balaton in western Hungary. It was a needed vacation, not least since he had just learned that he would soon travel to Sweden as part of a group of students. On the second day of the holiday, Eva joined the group. Georg recalled watching her arrival from a distance, unaware of who she was at first, only observing that she was 'a dark girl with a strange, breathtaking beauty'. On closer inspection, he 'perceived a most unusual combination of hilarity and sorrow, seriousness and play in her eyes'. Eva was still engaged to her long-term boyfriend but had doubts about the future of her relationship with him. Thoughts of him must have moved quickly to the back of her mind as she and Georg began an intense romance, at least according to Georg, who later wrote that they shared 'a passion that conquered my entire consciousness with the force of an elementary power'.[73] What Eva thought about their summer romance is more elusive. But, when Georg left the summer house to prepare to travel to Stockholm, she believed she would never see him again. Georg, on the other hand, was determined to see Eva again. After spending several weeks at the Cell Research Department at Karolinska Institute in Stockholm, Georg was offered a job at the department. The timing couldn't have been better, or worse. 'I still remember the mixture of ecstatic happiness and enormous anxiety,' he later wrote.[74]

> My situation appeared totally hopeless. I knew virtually nothing. I was halfway through my medical studies, still far removed from an MD. I was desperately in love with a girl whom I had only known during a summer vacation of eight days and who was on the other side of an increasingly forbidding political barrier. I did not know a word of Swedish. Still, I was firmly decided to resist the more comfortable possibility of continuing my studies in Hungary.[75]

The political climate in Hungary was quickly deteriorating as the country fell under the power of the Soviet Union through the Communist Hungarian Working People's Party. Aware that the Iron Curtain was falling on Hungary, Georg returned to Budapest in September 1947 with all the paperwork necessary to go back to Sweden, at least for himself. The problem was whether he could get Eva out of Hungary and to Sweden as well.[76] Because their reunion, they later wrote, 'confirmed what we both knew already: we wanted to live and work together', the first logical step was to marry secretly before Georg had to return to Sweden. After an intense day going from government office to government office making token bribes to officials to expedite the necessary paperwork, they said their vows with two friends as witnesses late on a Friday afternoon. 'Afterwards we ate our wedding dinner on the hall bench together with our witnesses,' they later recalled. 'There was only one dish: [Georg's] mother's carefully packed goose liver sandwiches. In the evening we went back to our parents' homes where no one suspected anything.' That Sunday, Georg returned to Sweden without Eva so he could arrange the necessary paperwork in Sweden. Finally, in March 1948, they were reunited in Sweden.[77]

* * *

Ruth and Alfred Pick had an advantage that most would-be immigrants to the United States did not have: jobs waiting for them. Dr Richard Langendorf, Alfred's former colleague from Prague, who escaped Czechoslovakia soon after the occupation of the country by Nazi Germany, had successfully established himself as a cardiologist at the Michael Reese Cardiovascular Institute in Chicago, Illinois, with help from the head of the Institute, Dr Louis N. Katz. These important connections helped to secure jobs for

Ruth and Alfred at Michael Reese, which in turn helped to support their visa applications to the US along with affidavits from family members in the country. A history of Michael Reese Hospital published in celebration of its centennial explains that the wartime assistance to physicians affected by the Nazi regime was extended into the aftermath on an informal basis to assist Ruth and Alfred. According to this account, Langendorf appealed to Katz, who 'sent an electrocardiogram from a patient with a puzzling cardiac irregularity' to Alfred. Alfred's correct diagnosis so impressed Katz that he immediately offered Alfred a paid fellowship and Ruth an unpaid position as a lab technician in the department.[78] All that remained was for the couple to secure the necessary paperwork and make their way to the US. But even with the bureaucratic support of the Joint Distribution Committee and financial assistance from family members in the US, the process dragged on due to the increasingly unstable political conditions in Czechoslovakia.

The Communist *coup d'*état in February 1948 marked the beginning of a forty-one-year totalitarian regime in Czechoslovakia. 'It is not surprising that large numbers of people should wish to escape from a country where spying and informing on a neighbour or workmate is declared a civic duty,' wrote one Briton who visited the country after the coup. So desperate were many to escape that they left everything they had behind and risked imprisonment if caught. Only limited numbers were allowed to leave, with the informant declaring that only 356 people had been permitted to leave the country in September 1949. 'As an additional safeguard,' the informant wrote, 'all passports became invalid on 30 November [1949], and they will probably be replaced by documents valid for very short periods only'.[79] Fortunately, Ruth and Alfred managed to leave the country before the situation reached this point. In late 1948, they received their transit visas, which allowed them to take a train to the port city of Gothenburg, Sweden. On 5 January 1949, they boarded the Swedish American Line ship the MS *Gripsholm* for the transatlantic journey to New York, where they arrived on 16 January. 'The first time we got peace and quiet was here,' Ruth later said about their arrival in America after the years of struggle and persecution.[80]

* * *

Victims of Nazi persecution who survived the war were granted the possibility of a new life in the aftermath of destruction. But these afterlives came with no guarantees. Surviving in the aftermath of the Second World War and the Holocaust was plagued with illness, loss, destruction, and political persecution, not to mention uncertainty about what the future might hold in terms of establishing new lives and careers. Many did not survive in the afterlife because they succumbed to illness and disease or to the psychological traumas that they had experienced. Still others lost their lives to repressive regimes or were trapped within them for decades. Halina Czajkowska, Halina Neujahr, Ellen Loeb, Ruth Pick, Rita Levi-Montalcini, and Eva Klein were among the lucky ones who survived not only the Holocaust and the Second World War but also the aftermath. As the 1940s ended, all six women found themselves in countries where they could live without fear of political repression and build new lives. New challenges now awaited them: new languages, new social customs and norms, new biases against foreigners and religious minorities, and – of course – a new era of science, medicine, and technology that had little place for women. All women faced the gender bias in STEM in the postwar era, but the six women in this book faced a host of other biases as well as they pursued education and careers in these male-dominated fields. But they had overcome worse and were not to be underestimated.

Chapter Five

Running Toward the Future

> We tried to forget the war, including the Holocaust, which killed many friends and relatives. We had miraculously survived, escaping the transports to the death camps in different ways, and making do with false papers. In Sweden, we decided to devote ourselves to our new lives, to the new country, to science, and to the children we were going to have…
>
> Eva Klein, MD and PhD, and Georg Klein, MD, 1993[1]

While deliberately forgetting the war and the Holocaust may have been a priority for many survivors of Nazi persecution and genocide, the past was never far away. The experiences of everyone who experienced these cataclysms inexorably changed their lives and outlooks. And, whether they acknowledged it or not, the careers of Halina Neujahr, Halina Czajkowska, Rita Levi-Montalcini, Eva Klein, Ellen Loeb, and Ruth Pick were altered and informed by their experiences during the war. Everything that came after was effectively part of the past even as it drove the future of medicine and science. In a macabre way, the achievements of each of these women was made both because and in defiance of the attempt to extinguish them.

At the same time, everything they accomplished was due to a combination of talent, determination, and good fortune. Had one or more of these elements been lacking, humanity would have lost out on their contributions. It is impossible to know how many medical and scientific discoveries might have been possible had more women encountered as relatively favourable circumstances as these did. That said, each of these women had to overcome significant barriers to achieve their positions in STEM and thus make their contributions to their respective areas of science, technology, and medicine. If we think of the various barriers as hurdles, women in STEM who were also immigrants in the countries where they studied and worked had to jump over more hurdles than women who dared to enter the competition in the countries in which they were born. Lapses in education and professional experience meant that additional hurdles had to be overcome to access further education and employment. But, of course, before these were the hurdles of language, social, and cultural barriers. That is assuming that they did not carry their past in a way that might make others feel burdened by it or affect their perception of them.

In short, nothing came easy to these women. But, far from running away from the past, their pasts seemed to be propelling them forward. Each seemed to be running almost feverishly towards the future, pursuing education and work as if it were a necessity of their existence.

Growing Up (Again)

Survivors of Nazi persecution like Halina Neujahr and Halina Czajkowska had been brought to Sweden in the spring and summer of 1945 through two humanitarian relief efforts. The first of these was the so-called 'White Buses' effort, led by the Swedish Red Cross. In early 1945, faced with a crumbling Third Reich, Heinrich Himmler – one of Hitler's most senior officers – went behind Hitler's back to secretly negotiate for the release of a relative handful of prisoners. His motives were purely self-serving. With Nazi Germany losing the war, Himmler – who, as the chief architect of the Final Solution, had the blood of as many as 14 million people on his hands – was using the lives of the prisoners as human currency to improve his prospects following the inevitable Nazi defeat. On the other side of the negotiations was Count Folke Bernadotte – a member of the Swedish royal family and vice-president of the Swedish Red Cross – who began secretly negotiating with Himmler in Berlin in February 1945 for the release of Danish and Norwegian prisoners. The scope of the agreement initially limited the Swedish expedition to relocating Scandinavian prisoners to a central concentration camp where the Red Cross would see to their care. But whatever advantage Himmler imagined that he would have quickly dissolved, and he agreed to Bernadotte's request to also liberate female prisoners of all nationalities and evacuate them to Sweden. In April and May, as many as 21,000 prisoners – mostly women – were taken out of active concentration camps and brought to Sweden in buses painted white with large red crosses on them.

The second effort took place in the summer of 1945, following the end of the war. It is often referred to as the 'White Boats' effort since it involved boats painted white with large red crosses carrying very ill prisoners – including Halina Neujahr and Halina Czajkowska – to Sweden for urgent medical treatment at the behest of the United Nations Relief and Rehabilitation Administration. Nearly 10,000 more survivors were brought to Sweden in this way.[2]

Initially, these individuals – who were referred to in government documents as 'repatriates' – were expected to stay in Sweden no longer than six months or until they were healthy enough to leave the country. Of course, some did want to return to their home countries, especially Danes and Norwegians. But many others were afraid to return to their homelands, often for religious and political reasons. Poles – who comprised the largest group of survivors in Sweden after Scandinavians – were among those who were afraid or unwilling to return. Initially, they had no choice in the matter. In October 1945, the first group of Poles to be repatriated were placed on a ship at the harbour in Malmö. Several jumped off the ship while it was still docked in Sweden, so desperate were they not to return to Poland.[3] Between 6 October 1945, and 12 January 1946, 5,102 Poles were repatriated, nearly 40 per cent of the approximately 13,000 who had come to Sweden as refugees in the spring and summer of 1945.[4] Halina Czajkowska and Halina Neujahr were still receiving medical treatment during that period, which was fortunate, as both women had been determined not to return to Poland even before they were brought to Sweden. Once Halina C. and her mother Emilia learned that Halina's father was alive and well, their choice was vindicated because of his involvement in the Polish underground organisation, the Home Army (or 'AK'). Halina C. later explained, 'Father wrote that he could not return to Poland, as he would be immediately arrested on arrival, having worked in the AK throughout the war. The Communists considered everyone who had worked in that organization as "Enemy No. 1."'[5] Halina Neujahr's decision to remain in Sweden had always been because there was nothing to return to in Poland, where her home and family were destroyed and antisemitism remained a significant problem.

In late winter 1946, new pressure was placed on the repatriates when a March 1946 government memorandum printed in Swedish and Polish informed the 'remaining Polish refugees' that the 'question has now become relevant about the return home during the early summer' of those 'who are unable to give any acceptable reason for remaining in Sweden'.[6] Fearing repatriation, Emilia immediately wrote to the Swedish immigration authorities 'asking most earnestly for the right of asylum for my daughter Halina and me here in Sweden'.[7] The request enabled the two women to remain in Sweden while an investigation was carried out which eventually granted them the status of political refugees.[8] Due to her continued poor state of health at the time of the March 1946 memorandum, Halina Neujahr was exempted from the immediate threat of repatriation, and did her subsequent studies at the Polish schools. Fortunately for Halina N. and Halina C., by the time they completed their basic education, they had gained the conditional right to remain in Sweden. But now, their continued presence was contingent on their ability to support and behave themselves, and their movements – even within the country – were limited and subject to approval.

* * *

During their first years in Sweden, both women had lived as young students – serious about their studies but also enjoying carefree social lives in protected environments. Having made up for lost time and completed their studies, by mid-1947, they were now fully-fledged adults, expected to fend for themselves, with the caveat that they had to do so in a country not their own, where the language and social customs were still unfamiliar to them.

As long as she was attending the Polish school in Sweden, learning with and being taught by other survivors of Nazi persecution, Halina Neujahr had been living in a comfort zone of sorts. After the Öreryd refugee camp closed in late autumn 1946, the Polish school was moved to a summer house in the small village of Vikingshill, Saltsjö-Boo, in the Stockholm archipelago. Halina N. with the other students and teachers, including Halina's favourite teacher, Piotr Eisenstadt, moved into the new location in November 1946, and picked up where they had left off in Öreryd. For the students, this meant making up not only for their lost education, but also their lost youth. Halina

N.'s personal photo albums are filled with snapshots of her and the other students looking healthy and happy while rowing, hiking, and playing in the snow. One would never know looking at the images that only a year or two earlier, they were mistreated prisoners facing a death sentence. During her year at Vikingshill school, Halina excelled in nearly all subjects, most notably maths, physics, and chemistry. Her completion certificate, dated 3 September 1947, notes, 'Halina Neujahr has been superior to her peers in terms of achieved study results.'[9] Nearly 23, Halina N. had defied the Nazis again by completing the education the regime had denied her. But, like Halina Czajkowska one year earlier, she now found herself in a new and uncertain position with no family support and an uncertain political status in Sweden.

After Halina Czajkowska left Öreryd in September 1946, the government sent her to live near the city of Falun, famous for its copper mines, in the north of Sweden. It was not where she wanted to be, but she made the best of it.

> Two hundred and fifty kilometers northwest of Stockholm, the new place was again a settlement for concentration camp survivors, who were foreigners of many different countries, some in good health, others not so well. Our group was provided with a study room, where we reviewed our matriculation courses so that we would not forget them. We also endeavoured to learn both Swedish and English, which was not an easy task.[10]

Her mother, who was living in Gothenburg, Sweden, visited her. Among their most urgent concerns was how to get Halina C.'s father to come to Sweden so the family could be reunited. More immediately, Halina C. hoped to settle in Stockholm, where she could begin to study for a career and find a job. Halina Neujahr had similar ambitions. The problem was that to gain permission to live in Stockholm, refugees first needed to secure a place to live in the city. As Sweden's capital and most populated city, Stockholm's population was around half what the Polish capital Warsaw's had been before the war.[11] Unlike Warsaw, which was destroyed during the war, Stockholm and the rest of Sweden had been spared due to the country's nominal neutrality during the conflict. Nonetheless, housing was a problem in the country, including in its biggest city. In the early twentieth century, Sweden's housing standards were among the worst in Europe, a reality that began to be addressed in earnest in the 1930s but stalled to an extent during the war. As in other countries, the postwar period came with a baby boom that made the housing crisis even more acute.[12] This meant that finding even the most modest accommodation was extremely difficult. Despite this and with little money, both women managed to find accommodation in the city by fall 1947. Halina C. secured a large room with two sofas, one of which was occupied after ten in the evening by the landlady's granddaughter. She considered it a temporary solution until she could secure a job that would pay her enough to rent something larger where she and her mother could live together.[13] Having secured a job as an office assistant at the Swedish postal service immediately after graduating, Halina N. stayed with a family for a month until she found a somewhat more private space to rent. She would live in rented rooms for a few months or years at a time for the next decade.[14]

It would have been understandable if Halina Neujahr and Halina Czajkowska, who had been through so much over the past eight years, took full advantage of being untethered young women living in a European capital and simply enjoyed themselves. Instead, they wasted no time taking the next step in their education.

Education Against All Odds

Almost immediately after securing her room in Stockholm, Halina Czajkowska registered in the building (architectural) engineering programme at Stockholm Technical Institute (STI), a private vocational school for engineering education founded in 1924. The demand for more and better housing in Sweden saw the demand for various types of mid-level engineers rise rapidly from the 1920s to 1940s. The student body at STI increased from just over 2,000 in 1939 to around 3,200 in 1946, in part due to the large number of foreign students admitted. The school's newsletter proclaimed in February 1949, 'A significant part of the students, namely 475, are foreigners, representing 19 nations. 217 Norwegians make up the largest group, followed by 95 Estonians, 53 Finns, 52 Poles, 12 Icelanders, 11 Danes, and others.'[15] In the autumn 1947 semester, when Halina C. began her studies at STI, she was one of around forty-four Poles enrolled, making them the largest single national group at the school after Swedes and Estonians. Very few of the students were women.[16] During Halina's time at STI, students could study subjects like mechanical engineering, building (architectural) engineering, electrical engineering, road and water engineering, chemical engineering, and so forth. Depending on factors like previous education and the type of certificate or diploma sought, students studied for between half a year and two and a half years.[17]

Exactly why Halina C. chose to attend STI is not clear, but it was probably due at least in part to the fact that foreigners could register as regular students with a path toward a diploma, something that was not possible at other institutions. Hence the large number of Poles and other foreign students enrolled there. In addition, the cost of tuition was also relatively low, at least according to the school's principal.[18] STI also offered the option of day or evening classes, which made it possible to work and study if necessary. With the help of a small scholarship to pay for tuition and rent, Halina initially took day classes, though she was left with 'barely [...] enough to pay for adequate meals'.[19] Not that she had much time to eat or socialise. During her first semester, she took classes in mathematics, trigonometry, planimetry, slide rule, mechanics, strength of materials, physics, general chemistry, materials science, and German language. Her grades varied from fair to excellent. Not surprisingly given her recent experience, she achieved top marks in German.[20] Swedish caused problems, however, as she recalled in her memoirs:

> With my extremely limited knowledge of the Swedish language, it was very difficult. My salvation lay in the fact that I was good at mathematics and anything to do with numbers, and also at drawing and drafting – subjects that did not require exquisite language skills. But, to prepare for the classes, I often had to work until 2 a.m., translating paragraphs in the textbooks word by word (from Swedish to Polish), so that I could understand the next morning's lectures.[21]

Just as she was about to complete her first gruelling semester, she received a telephone call informing her that her mother was in the hospital. Mother and daughter had been separated when Halina C. left for the Polish school in Öreryd in early January 1946, leaving Emilia behind in Gothenburg, where she worked in a chocolate factory. While Halina was thriving, Emilia was struggling. They had only seen each other during short visits since their separation, and Emilia was suffering from loneliness, among other difficulties. While Halina was in Öreryd, Emilia's letters to her 'became more and more depressed. She would often mention her experiences in the concentration camps, finishing with the word, "Why?"'[22] Although she was now in regular communication with her husband, Halina's father, who was in a displaced persons camp in the American Zone of Germany, Emilia – as Halina later understood – was almost certainly suffering from what we now call Post-Traumatic Stress Disorder (PTSD), a condition not fully understood or recognised until many decades later. In the postwar period, survivors' psychological trauma often went overlooked or ignored by humanitarian aid organisations and medical professionals. 'It seems altogether incredible today that when the first plans for the rehabilitation of Europe's surviving Jews were outlined, the psychiatric aspect of the problem was overlooked entirely,' wrote psychiatrist Dr Paul Friedman in 1948.[23]

After receiving the telephone call, Halina C. dropped everything – including her studies – to be by her mother's side at the hospital in Gothenburg, where the doctors were struggling to help her due to their inability to communicate with her. Emilia insisted that Halina bring her back to live with her in Stockholm as soon as possible, and this is exactly what Halina did. This meant that Halina had to stop studying at STI so she could find a job that would support them both. Soon, she and Emilia settled into new accommodation in Stockholm, and Halina was taking whatever work she could to support them. When no work could be found, she applied for financial assistance that carried them through. To add to her load, she was doing everything she could to bring her father to Sweden and was finally successful in December 1948. At just 21 years old, she was taking responsibility not just for herself but also both her parents. Despite everything, she re-enrolled at STI for the spring 1948 semester, this time taking evening courses, which enabled her to work during the day, and changing her area of study from building engineering to chemical engineering.[24] As before, she took on a full course load and achieved good grades her first semester. She continued in the fall semester, raising her grades further in most subjects, and continued to do so every semester until her last in spring 1950.[25] On 24 May 1950, she successfully passed her chemical engineering exams.[26] Of the twenty students in the evening chemical engineering programme, she was one of only three women. All three of the new female graduates were survivors of Nazi persecution in Sweden as refugees.[27]

* * *

Halina Neujahr could have enrolled at STI as well. She was almost certainly acquainted with at least some of the Polish survivors who did – including, possibly, Halina Czajkowska. As Halina N. later recalled, 'My colleagues – my Catholic Polish colleagues – all wanted to become engineers.'[28] Instead, Halina set her sights on studying chemistry at *Kungliga Tekniska högskolan* (the Royal Institute of Technology), known as KTH, in Stockholm. Established in 1827, state-owned KTH is the oldest technical university in Sweden. Prior to 1921, women could not study for degrees at KTH, and doctorate degree programmes were not offered until 1927. Through some of her Polish colleagues, Halina learned that, unlike STI, foreigners were not admitted as 'regular' students with the possibility of earning a degree at KTH. There was, however, the possibility for a limited number of foreigners who had already studied in Sweden to be admitted as 'special' or 'extra' students who, if they passed the necessary examinations, could apply to the government to become a regular student and earn a degree. Confident in her grades, Halina N. decided to go directly to the rector (principal) of KTH, Dr Ragnar Woxén, who held an open office every Monday afternoon for one hour, and ask to become an extra student. 'I had to ask my foreman at the post office that I would like to take off a Monday afternoon. And she asked me – it was a woman, very strict – "What for?"' Halina recalled, continuing with a laugh. 'So, I said, "Because I have to meet the rector of the Royal Institute of Technology." And she looked at me as if I were out of my mind, but she didn't say anything, and I got the time off.'[29] The story of Halina's visit to Woxén's office was described in almost mythical terms in a 1993 article in the KTH newsletter:

> One spring day in 1947, a young Polish woman entered KTH rector Ragnar Woxén's room. [...] The young lady surprised in several respects. Dressed in

old surplus clothes, Halina Neujahr showed an exceptionally good student grade and explained in stuttering Swedish that she wanted to start studying at KTH.[30]

Mythologising is understandable but unnecessary. Halina was determined to study and had the qualifications to do so. The only thing standing in her way was her nationality. That said, it was probably fortunate that Woxén was the rector of KTH at the time. Woxén (1905–1994) graduated from KTH as a civil engineer in 1928 alongside Greta Westberg (1902–1993), the first woman in Sweden to become a civil engineer. The two married the following year. Woxén was almost certainly progressive when it came to women in engineering and Halina's grades clearly impressed him. 'So, he said, "Oh, but I have never seen such a good diploma,"' Halina N. recalled. '"Usually, we don't take in people who haven't started studies elsewhere, but I shall do my best,"' he told her. Woxén advised Halina about the steps she should take and put her in touch with various people at KTH. By early 1948, she had secured the practical experience in chemistry that was necessary before she could be accepted as an extra student at KTH. On 8 February 1946, she left her job at the postal service and began working as a laboratory assistant at *Svenska Träforskningsinstitutet* (the Swedish Wood Research Institute, or STFI) on 9 February 1948. Her persistence paid off when, on 1 October 1948, she began studying at KTH as an extra student in chemical engineering, taking two courses and working as a technologist. 'And then I finally found myself as a single girl amongst thirty [male] students of the first year, which was very surprising. But I didn't mind.' There was more on her mind, in any case. Such as twelve hours of mathematics per week and the fact that she had been admitted to KTH on approval. Only high grades and passing grades on the necessary exams would enable her to petition for the right to earn a degree.[31]

'I was a lonely girl, and everyone was very kind to me,' Halina N. recalled. In addition to being very alone, like Halina Czajkowska, Halina Neujahr's biggest challenge in the beginning was the language barrier. 'The first year was terrible because I didn't understand what they [the professors] said, I didn't understand anything,' she explained. 'It was only when I had acquired some English-language books, e.g., in physical chemistry, that I understood what it was really about.'[32] Though she stood out in many ways among her fellow students – not least as a foreign woman – she never discussed her past with them. She considered her war experiences a 'humiliating' topic that they wouldn't or couldn't understand anyway. 'I came to another world. Those people didn't have any idea of what it [Nazi persecution during the war] was… especially not those young people, the students. Not even the professors.'[33]

After an intensive first year of study, Halina N. spent the summer holidays working again at STFI before starting a new school year on 1 September 1949. Exactly ten years earlier, she had been due to return to school in Warsaw but was prevented from doing so by the Nazi invasion of Poland. Now, she was entering her critical second year at KTH. As before, she supported herself by working as a technologist at KTH and taking summer jobs – gaining practical experience at the same time she was studying – while also relying on student loans from the Stockholm Jewish Community.[34] Following the completion of her second year, Halina applied for and was granted a place as a regular or 'ordinary' student at KTH and continued her studies in earnest as she worked toward her degree in civil engineering. As important as gaining this right was, the research was as important – and possibly more so to Halina – than the degree. Originally, she should have graduated in February 1953, but she postponed this when offered 'the opportunity to carry out interesting research work in direct connection with my studies and in an area that I also intend to pursue in the future', she wrote at the time.[35] This assignment was her introduction to researching the newly discovered vitamin B12, which would indeed define her future scientific career. In addition, between August and October 1952, she had an internship in France that gave her further practical experience. Finally, after years of hard work, Halina N. earned her civil engineering degree in chemistry and became a Swedish citizen, both in July 1953. At the time, she owed around 10,000 Swedish crowns in student loans to the Stockholm Jewish Community, and another 1,500 Swedish crowns to the Committee for Stateless Refugees in Sweden, the equivalent of around 20,000 Euros today.[36] In debt, she eagerly accepted a job at a candle company, but it was not to be.

* * *

Twenty-three-year-old Eva Klein's early experiences and challenges in Sweden were different in many ways from those of Halina Neujahr and Halina Czajkowska. Housing and employment were already arranged for her by the time Eva arrived in Stockholm in March 1948. She and Georg settled into a rented room and Eva began working as a research assistant alongside Georg at the Karolinska Institute (KI) in Stockholm immediately after arriving. The housing was miserable, according to Eva – a furnished apartment but without a toilet or kitchen.[37] But at least she was welcomed with open arms by Professor Torbjörn Caspersson, head of the cell research department at KI, who had done the same for Georg the previous year.[38] Established in 1810 to train army surgeons and granted a licence to confer medical degrees in 1874, KI is Sweden's third-oldest medical school and holds the right, per Alfred Nobel's will, to select the Nobel laureates in Physiology or Medicine. Women studied and earned degrees at KI in the late nineteenth century, though it was not until 1900 that a woman earned a doctoral degree. It took nearly forty more years before the first female professor, Nanna Svartz (1890–1986), was appointed in 1937, a position she held from 1938 to 1957.[39] In the mid-twentieth century, gender inequalities were not as prevalent in medical education and work in Sweden as they were in chemical engineering, but they still existed. Women comprised only around one-fifth of all medical students at KI, and Svartz was the only female professor until her retirement in 1957, leaving a void that was filled again only in 1965.[40]

Gender inequality was likely the least of Eva's concerns, however. Both she and Georg wanted to continue to study medicine and complete their medical degrees, preferably in Sweden. But, with passports valid for only one year without the chance of an extension, this was impossible. At the same time, they were loath to return to Hungary, which was under the thumb of ardent Communist Mátyás Rákosi, whose repressive regime and deadly political purges were patterned after those of his role model, Soviet leader Joseph Stalin.[41] Though Rákosi was Jewish, he identified only as a Communist, and 'did not hesitate to ride the waves of antisemitic public sentiment'.[42] The couple's desire not to return to the country was a professional imperative as well. Georg understood this already during his first stay in Sweden in 1947, when he realised, 'If you were young and wanted to have a future in science, you should get your degree as soon as possible and leave the country [Hungary].'[43] Any doubts he and Eva held that this was still the case were dispelled in August 1948, when they attended the Eighth International Congress of Genetics in Stockholm. Addressing the assembled scientists in his role as president of the Congress, American geneticist and Nobel Laureate Hermann Joseph Muller condemned Soviet agronomist Trofim Lysenko for declaring that the scientific study of genetics was a 'bourgeois perversion', a claim supported by Stalin with the result that the study of genetics in the Soviet Union was effectively banned.[44] It was not just the study of genetics that was imperilled, Muller told the assembled researchers, including Eva and Georg, but also geneticists like them.

> Most of the main geneticists [in the Soviet Union] are, as the case may be, broken, executed, disgraced, not to be located, fugitives, forced into other work or, in the mildest cases, driven to such a redirection of their lines of research and publication as to give the appearance of working somehow in support of the doctrines approved by the officials.[45]

Georg and Eva were all too aware that if science and researchers were under threat in the Soviet Union, they were also under threat in Communist countries like Hungary. As desperate as their situation was, they held passports, which meant they could not claim refugee status that would enable them to remain in Sweden and study medicine. 'To get into medical school here [in Sweden] it was very difficult. They took refugees from the Balticum [countries in the Baltic region], but we couldn't say we were political refugees as long as we had the passports. So, it was a Catch-22,' Eva explained. Fortunately, Caspersson intervened. Before their passports expired, Georg was admitted to medical school at KI, ensuring the couple could remain in Sweden. No provision was made for Eva beyond the continuation of her work at KI. Studying for her medical degree would have to wait until Georg completed his. On reflection at least, Eva was pragmatic about this, emphasising her role as a breadwinner while Georg studied.[46] After Georg completed his medical degree in 1951 and received the position of assistant professor of cell research at KI, it was Eva's turn. In the meantime, the couple had also started a family, and Eva juggled her medical studies with motherhood, as well as Georg's feelings about her roles as mother and scientist.

> According to Eva, from the time their children were born George [sic] never participated in any housework. When the first child, their son, was born, she invited to Stockholm the nanny who raised her in Budapest to live with them in Stockholm. Later, too, they always had live-in and other help, but 'still it was a disaster.' What made it even more difficult for her was that George not only did not do anything, but on top of that, he hated that she was involved with 'nonscientific' activities. Naturally, this made it even harder for Eva.[47]

Eva persisted, earning her medical degree in 1953 and her PhD in tumour biology in 1955. At the urging of Caspersson, she defended her thesis while she was eight months pregnant with her second child. By then, her groundbreaking research was already making headlines, and she received the position of associate professor at KI.

* * *

On 1 July 1943, two hundred students began studying medicine at the newly established Southwestern Medical School in Dallas, Texas, located around two miles from downtown Dallas and less than half a mile from where Ellen Loeb's sister Trudy Shakno lived with her family. Assembled in two months during wartime rationing, the new medical school was primitive, composed of prefabricated buildings that were described as 'shacks' and surrounded by 'grasslands and a few roads'.[48] Some of the young men studying at Southwestern were Army or Navy recruits. Nonetheless, they were completely disconnected from the reality of the horrors of the war and the Nazi atrocities. At the same time Ellen was fighting for her life and those of her young patients in Westerbork, the medical students in Dallas were experiencing, as one described it later, 'the turmoil, the discomfort of the heat of that first summer in 1943; the scurrying and the crowding and the noise of our barrack buildings; the incredible stench and depressing atmosphere of old Parkland hospital'.[49] Though Ellen had survived far worse conditions, the thought of returning to anything remotely reminiscent of her time in the concentration camps might have repelled her. Instead, in late 1946, 26-year-old Ellen set her sights on attending Southwestern. Although it was just a short walk from where she and her mother were living with Trudy's family, entering that medical school was an arduous trek.

Before she left the Netherlands, Ellen secured documents attesting to her education and work as a nurse. One of the documents, dated 4 January 1946, describes her as a medical student in Amsterdam, though it does not elaborate on the specifics of her education.[50] Another was written by Dr Salomon Kroonenberg (1891–1961), medical director of *Joodsche Invalide*, a Jewish nursing home in Amsterdam. Kroonenberg had been deported to Westerbork in 1943, where he worked as a paediatrician and knew Ellen.[51] He attested that, as a nurse, Ellen 'devoted herself to her task with great dedication and tact and thus achieved good results, which is why he can especially recommend her for childcare'.[52] While these documents would have been valuable to her after she and Dina arrived in Dallas in the fall of 1946, they alone were not going to get her into medical school.

Southwestern Medical College required prospective students to apply one year in advance and accepted a limited number of first-year students to begin studying in the autumn term each year. Entrance requirements were high, according to a handbook published in 1948, which stated that consideration 'will be given to those with a minimum of three years or 90 semester hours of work (not including required physical education or military science) from an approved college of arts and sciences'. Of the ninety semester hours of work, the prospective student was required to have twenty hours of chemistry, eight hours each of physics and biology, and one academic year of English. A bachelor's degree was recommended, along with one year of college mathematics and 'one or more modern foreign languages'. Those who met these requirements would then be evaluated based on any pre-medical training, a medical aptitude test, and a personal interview. The application fee was 5 US dollars (USD), the equivalent to around 66 USD in 2025, and, if accepted, tuition of 175 USD (around 2,322 USD in 2025) for the first semester was payable within two weeks of acceptance.[53]

Ellen wasted no time. Almost immediately after arriving in Dallas, she began studying English at night school and started working as a nurse part-time, possibly at nearby Parkland Hospital.[54] Assuming Ellen applied one year in advance, she would have been among the record 617 applications for admission made in 1947, only 64 of whom were accepted.[55] She was among the successful applicants. 'Because she knew what she wanted, that impressed the dean,' Trudy explained. To pay for the tuition, Dina sold some of the jewellery, furniture, and other valuables that she and Ellen had successfully recovered in Amsterdam.[56] In the fall of 1948, barely two years after arriving in the US, 28-year-old Ellen began her first semester at Southwestern Medical College as one of only three women in her class.

The 1949 student yearbook is a reminder that the students at Southwestern were almost exclusively white and mostly male. In Ellen's class, just one student was Hispanic, and he was from the US territory of Puerto Rico. Like other southern states, Texas held tight to racial 'separate but equal' segregation policies that prohibited or limited Black and Mexican Americans' access to public education (including higher education), transportation, and spaces. Southwestern was no exception. Had there been any Black students, they would have been horrified at the crude and racists depictions of Blacks in the yearbook.[57] What Ellen thought of these racist images and how American racism affected her personally are unknown. It is also unclear whether she divulged personal information to her fellow students, including that she was Jewish and had survived the Nazi genocide of Europe's Jews. Dallas had a relatively large Jewish population in 1950, roughly 10,000 of the approximately 434,000 residents of the city. In Texas, only Houston, which was more populated than Dallas, had a larger Jewish population – 14,000.[58]

While Jews were unwelcome in many private and some public spaces, they were not prevented from public education as were Blacks and Mexicans. Ellen was not the only Jewish student at Southwestern – she was not even the only Holocaust survivor – and the school probably also had one or more Jewish professors. There were also other Germans in Dallas, who, like Ellen's sister Trudy and her family, found refuge there in the 1930s.[59] But this did not mean that she felt comfortable – or even safe – disclosing or sharing certain aspects of her identity and experience with her classmates. Regardless, she was socially active. The Shakno family had moved to a house slightly further away from Southwestern, and Ellen and Dina continued to live with them. Although she lived off campus, Ellen joined Alpha Epsilon Iota, a professional sorority for women in medicine, which would have provided her with the support and camaraderie of a slightly larger group of women. She features in several posed and candid photos in the 1949 student yearbook, looking relaxed and comfortable with her fellow students, male and female. In two photos, she is referred to endearingly as 'Mother Loeb'.[60] Ellen was not the only motherly figure at Southwestern. According to Trudy, Dina took a job washing dishes at the school to help pay for Ellen's tuition and expenses. 'She did that for about four months. And then one day, the cook didn't show up in the cafeteria, the school cafeteria, and she cooked.' The students liked her cooking so much she was hired as the cook, first at the school and then at Alpha Epsilon Iota.[61]

The first-year student course load at Southwestern amounted to over 1,200 hours, not including study time. This increased to nearly 1,450 hours in the second year and peaked in the fourth and final year at 1,620 hours.[62] In addition to a heavy workload and busy student life, Ellen also served as a class officer, holding the role of Secretary and Treasurer at the time of her graduation.[63] Structurally, the conditions in which she and the other students studied had changed little since the school was established. Lessons were held in the same prefabricated 'shacks' that were put up in 1943. Driving to the school for the first time, one new medical student passed by it in his car several times, mistaking it for 'some sort of junk heap'.[64] The shacks had no air conditioning, which must have made for stifling conditions in the warm months, especially when the students had to perform autopsies on cadavers. Then there was the shacks' proximity to train tracks. The passing freight trains were so loud that lectures had to be paused until they passed. Despite Southwestern's primitive appearance and conditions, it grew in stature in September 1949, when it became part of the University of Texas system and was renamed Southwestern Medical School of the University of Texas. Despite this, the shacks were not replaced until after Ellen graduated.[65]

On 1 March 1952, Ellen was naturalised as a US citizen. The following day, a photograph of her and three other new citizens at the naturalisation ceremony appeared in a Dallas daily newspaper. One of them was her classmate and fellow Holocaust survivor, Larry Gladstone (1910–2010), who was originally from Czechoslovakia.[66] Almost exactly three months later, after four years of intensive study, Ellen received her Doctor of Medicine degree on 2 June 1952, as did Larry Gladstone. She was one of seventy-two graduates that year, only five of whom were women.[67] At the time of her graduation, she already had an internship lined up at the Veterans Administration (VA) Hospital in McKinney, an agricultural town with a population of around 10,000 located roughly thirty miles north of Dallas. Ten years earlier, Ellen's medical career began in Westerbork as a means of survival. During that period, she had her first encounter with the treatment of blood diseases while working as a forced medical labourer with German doctors. Now, she was a US citizen effectively starting her medical career all over again, only this time as a qualified physician.

Staying with the Trouble

Throughout what she called 'the difficult years' of the Fascist regime in Italy and the Second World War, Rita's actions – especially those related to her research – were driven by a combination of drive, passion, focus, and tenacity. In the face of every adversity, scarcity, and danger, she not only adapted, she thrived. Whether by painstakingly sharpening needles to make minute scalpels, hunting down fertilised chicken eggs for analysis, or protecting her equipment and samples during air raids as if they were her children, she also demonstrated an artist's creativity and a musician's ability to improvise. These traits were as important to her career success and longevity as were her formal education and training, and not only during the years she dubbed 'difficult'. In different ways, all years were difficult. The depression that beset Rita soon after the liberation lingered in the early postwar years and was soon joined by feelings of inadequacy and pessimism about the future of her research and the burgeoning field of experimental neuroembryology.

Though she made the journey by ship from Genoa to New York in the autumn of 1947, she did so filled with doubts. Her early encounters with antisemitism and Fascism in St Louis, Missouri did little to cheer her.[68] These experiences combined with advice from others led to Rita combining her surname since birth, Levi, with her mother's maiden surname, Montalcini, resulting in the hyphenated Levi-Montalcini. According to the anthropologist Ruby Rohrlich, 'The advice to change her name was Rita's early encounter with the attempt by American Jews to circumvent antisemitism by concealing a name that revealed their heritage.'[69]

Her first meeting with Viktor Hamburger at Washington University in St Louis, Missouri gave her a glimmer of hope which, together with her 'strong fascination' with the nervous system, kept her from abandoning research but did not ignite her enthusiasm.[70] Whereas Rita's depression would be treated today with a combination of antidepressants and therapy, little attention was given in the early postwar period to the psychological effects and treatment of wartime trauma. Rita made her own attempt at talk therapy by discussing her worries with her Italian colleagues, virologist Renato Dulbecco (1914–2012) and microbiologist Salvador Luria (1912–1991), who were living and working in Indiana. As was the pattern throughout her life, she found her best therapy in the laboratory. Late in 1947, while looking through her microscope at Washington University, she 'regained the faith in and the enthusiasm for the possibilities of experimental neuroembryology that I thought irretrievably lost'.[71]

It is possible, perhaps even likely, that other researchers had already seen what Rita saw that day in St Louis, which she later described as nerve cells 'acquiring a personality not usually attributed to them'.[72] However, her encounter with Giuseppe Levi three years later demonstrated that other researchers did not always greet anomalies with the same enthusiasm that Rita did. Even she described the prospects ahead of her as tenuous.[73] Had Rita been of a mindset that rejected the uncertain and anomalous and/or feared tackling findings that seemed to defy reason, logic, and rationale, she would almost certainly have ignored what she saw, as others may have done before her. Instead, her sense of curiosity and faith in her instinct propelled her to, in the words of gender philosopher Donna Haraway, stay with the trouble.[74] Depressed and pessimistic, Rita kept going – to the US, to Washington University, to the laboratory – and never stopped pursuing difficult questions about her research and the research field. It was a quality she shared with the other women in this book, each of whom was open to engaging in and pursuing 'unexpected collaborations and combinations', not only in human interactions but also in the type of human/non-human interactions that comprise much medical and scientific research, such as the work that takes place in the laboratory.[75]

It was by staying with the 'trouble' – persisting in the face of the difficult and unexpected aspects not only of their research but also their careers as women in STEM – that they made the discoveries that have contributed to humanity. The process was far from instantaneous. As Rita later recalled, what she found under the microscope that day in 1947 was not her most important discovery, and it took months of intense work to make sense of what she had found. But it marked what she described as 'the sealing of a lifelong alliance between me and the nervous system, an alliance I have never broken or regretted keeping'.[76] The commitment led to her remaining in the US and becoming a naturalised citizen in 1956 and, of course, her discovery of NGF.

* * *

Halina Neujahr's fascination with the 'beautiful red crystals' of vitamin B12 formed the basis of her own lifelong alliance with biochemistry. With the encouragement of her undergraduate supervisor, Professor Harry Lundin, she made the decision in mid-1953 to decline a secure job as a laboratory manager and instead continue her research as a postgraduate student at KTH in Stockholm. In doing so, she chose sewage sludge over candles and gave up a good salary that would have helped her pay off her student loans for a scholarship. Halina later insisted that when she resumed her research on vitamin B12, she had no plans to pursue a doctorate degree. Instead of a degree, it was her love for her research that seemed to propel her forward. She revelled in determining and mapping the organisms in wastewater that caused the synthesis of B12. Whether she had intended to or not, her continued research led to her defending her doctoral thesis – a six-hour ordeal that ended with her automatically becoming an associate professor and, of course, earning the distinction of becoming the first woman in Sweden with a Doctor of Technology (D.Tech) degree in chemistry.[77] She was also only the third woman in Sweden to earn this advanced degree in any specialisation, but for all the publicity she received, she might as well have been the first.

The Swedish press was quick to learn about Halina's experiences during the Second World War and as a refugee in Sweden and justifiably hailed her accomplishment as all the greater because of this past. Women's magazines were especially attuned to the layered and intersecting disadvantages and challenges she had to overcome. As one journalist wrote in 1963, 'with her talent and indomitable will as her only capital, [she] has created an international position in the world of technology'.[78] At the time of the interview, Halina had just returned to Sweden after spending eighteen months in the US, where she worked at the University of Illinois. There, she had a respite from sewage sludge, working instead in research related to another B-vitamin, biotin. Her work in the US had other advantages, including that she 'came into contact with the great scientific world and many famous biochemists' and that the 'range of attractive jobs at universities in the USA was very large in the "golden 60s"', she later explained.[79] But, once again, she turned away from the ostensibly 'easy' path in her life and work. Back at KTH in Stockholm, Lundin was working to change the department's subject area from food chemistry to biochemistry and asked her to come back and build the new department, virtually from scratch.[80] She accepted and returned to KTH in mid-1962, where she remained for the rest of her long career.

The same journalist who praised Halina in 1963 also powerfully yet elegantly summed up the connection between her experiences during the Second World War and the Holocaust and her research career when he wrote:

> Extracting life-giving and life-enhancing substances from the sewage network is in some way a parallel to her own life. Not even the deepest slump in human existence, such as Halina Neujahr had to experience during the years of horrors in the Warsaw Ghetto and Hitler's death camps, could defeat the positive core, the indestructibility of the intellect, which she has given proof of through her development.[81]

* * *

In the Theresienstadt Ghetto, Ellen had been introduced to the treatment of blood diseases and disorders when she was involved with the *heilgas* experiments for infected wounds. This and her experiences as a paediatric nurse in Westerbork shaped the course of her medical career. After completing her internship at the VA Hospital in McKinney, Texas, Ellen returned to Dallas to do her internal medicine residency at the VA Hospital there. At what point she chose to specialise in haematology is unclear, but it was not necessarily the most obvious choice for a medical resident. Even today, haematology is a demanding and emotionally challenging specialisation. 'In haematology, you are likely to encounter a lot of pain and tragedy – e.g. horrible aggressive haematological disease and a lot of deaths on the ward (acute leukaemia has a high mortality),' advises the National Health Service (NHS) in the United Kingdom. 'Seeing such things can be emotionally draining. And that's doubly true when patients die before you can save them.'[82] In Ellen's time, the mortality rate was even higher for blood disorders and diseases, and 100 per cent fatal for patients with leukaemia. Considering Ellen's experiences as a nurse in the Nazi camps, she might have been expected to choose a specialisation with better patient outcomes and fewer emotional demands. Instead, her medical experiences during the Holocaust taught her that in the face of death, she could improve her patients' quality of life, no matter how short. She could also contribute to medical research that increased patients' longevity, improved their quality of life, and contributed to finding better treatments and, perhaps, even cures for blood diseases and disorders. While the Second World War and the Holocaust had taken much away from Ellen, her experiences during that period had also profoundly shaped her personal ethics as a medical professional in the aftermath.

In 1952, the same year Ellen graduated from medical school, the J. K. Wadley and Susie L. Wadley Research Institute and Blood Bank opened in Dallas. Nine years earlier, while Ellen was in Westerbork transit camp, the Wadleys' only grandchild died of leukaemia. This tragedy inspired J. K. Wadley to make a substantial donation for a research institute in Dallas dedicated to treating and researching blood diseases and disorders. Ellen joined the Wadley institute in 1956 and worked as the clinical partner of the director, Dr Joseph M. Hill. As during the Holocaust, many of her patients were children and most were doomed to death, only this time due to incurable blood diseases.[83] Ellen dedicated the rest of her life to this work at Wadley. Although she never spoke or wrote publicly about her Holocaust experiences, they played a significant role in her clinical practice, as evidenced by stories left behind by her patients. In 1958, a woman with leukaemia who had been a patient at Wadley for two years wrote a letter to her local newspaper about her experiences and the physicians who treated her at Wadley. She found hope, she wrote, 'in the brown-eyed competent smile of Dr. Ellen Loeb, she, having survived three tortuous years in an [sic] European concentration camp, knows the importance of a determination to stay alive'.[84] After her death, Ellen's Holocaust story even made it to the US House of Representatives, when one of her former patients – a haemophiliac – opened his testimony in support of the Ricky Ray Hemophilia Relief Fund Act of 1995 with his recollections of how she and her experiences had inspired him.[85]

* * *

For a brief time, Eva Klein's and Halina Czajkowska's paths crossed at KI in Stockholm. Halina worked various jobs while studying chemical engineering at STI in Stockholm, including as a laboratory assistant at a brewery, *Münchenbryggeriet*, in Stockholm.[86] But, according to medical historian Jacalyn Duffin, Halina's 'biggest break and best-paid position came with her employment as a chemist in the Cancer Research Laboratory at the prestigious Karolinska Institute'.[87] In mid-October 1950, Halina left the brewery job with a salary of 350 Swedish kroner per month and took up her new position at Karolinska with a salary of 500 kronor per month (around 750 and 1,070 Euros in 2025, respectively).[88] In the position, she worked with the colony of inbred mice that Georg Klein had brought back from the US, thus contributing to his and Eva's groundbreaking research on ascites tumours that advanced the field of cancer immunology. In the process, as Duffin explains, Halina 'learned many of the histochemical and biochemical techniques, which were to serve her well later in Canada'.[89] Her fear of losing these specialised skills led her to discover the potential of vinca alkaloids in treating cancer in Canada. Her brief time at KI thus contributed to both cancer immunology research and the advancement of cancer treatment. Although Halina would have had direct interaction with both Eva and Georg in the laboratory, no known accounts of these encounters exist. However, when Halina decided to emigrate to Canada, Georg wrote her a letter of recommendation that helped her to secure the position at the Collip Medical Research Laboratory at the University of Western Ontario in May 1951.

Eva and Georg had also planned to stay in Sweden only until they completed their medical degrees. But the 'very important problem' of cancer immunology, into which they entered at Karolinska in Stockholm, kept them there for the rest of their careers, and they became naturalised Swedish citizens in the 1950s. Eva's style of research was the key to her scientific discoveries. 'Eva Klein's associative powers were also extraordinary,' according to one account. 'As a researcher, she saw details that others did not, and she often chose to investigate questions that the rest of the research world left untouched.'[90] While Georg preferred to move from one area to another, Eva dug deep into the same area, to 'look at the same old questions from different perspectives'.[91] This quality led to several major contributions to immunology, including the discovery of natural killer (NK) cells, which kill cancer and other infected cells, and the role of Epstein-Barr virus in Burkitt lymphoma.[92]

While their paths diverged geographically, both Eva and Halina C. stayed the course in their research paths despite their own and others' doubts. Halina made her most significant scientific contribution due to her determination to maintain her skills in her new position at the Collip Laboratory. Though it was fear of losing her specialised skills that drove her to conduct the white blood cell counts on the rats, it was courage that compelled her to risk her job and share her significant findings (conducted without permission) with her boss, Noble. Had she kept them to herself, an important breakthrough in cancer treatment would have been delayed or even lost to science. As for Eva, having overcome her own early scepticism about cancer immunology, she then maintained her course against the lingering scepticism of others. As a result, she not only broke important new ground in cancer research, she and her research also remain relevant today.

* * *

When Ruth began working at the Michael Reese Hospital in Chicago in 1949 as an unpaid lab technician, she was there primarily because her husband had been offered a position. Soon, she proved to her employer that she was worthy of a position based on her own merits, not just as the wife of Alfred Pick. While this led to paid work, it did not necessarily lead to professional equality with or respect from her male colleagues. To begin with, although Ruth felt that she was 'rich' when she finally received a salary, the reality was that she was paid a fraction of the amount paid to her male colleague, Dr Jeremiah Stamler, who had joined the department only one year earlier than Ruth. Ruth was paid a starting salary of fifty dollars a month as a technician (later increased to one hundred), but Stamler was paid two hundred dollars a month when he began working at Michael Reese in 1948.[93] Not only was Ruth paid less than Stamler, he also initially treated her as an intellectual and professional inferior. Dismissed not once but twice by Stamler while he was leading the atherosclerosis team, Ruth worked even harder to overcome his doubts about her and her abilities as a medical researcher, not to mention her important findings about the role of oestrogen in preventing atherosclerosis.

Ruth's unshakeable tenacity served her well at this critical moment as well as in her long career at Michael Reese just as it had in the Nazi camps. Those who doubted her did not know how hard-won this trait was. Her early experiences at Michael Reese were nothing compared to the terrible rite of passage she endured at Theresienstadt, where before she could work as a physician, she had to clean the filthy infirmary and deliver clean laundry. Even before that, she was effectively demoted from physician to nurse in the Jewish hospital in Prague during the Nazi occupation. These and so many other experiences during the war and the Holocaust were more than humiliating, they were part of a genocidal ambition designed to end with Ruth's death. Having miraculously survived, Ruth was not about to let the kind of setbacks and frustrations that she experienced in her postwar career in the US stand in her way. And she did not. In 1958, Ruth succeeded the man who once dismissed her research findings as a 'sampling error' as director of the atherosclerosis research project.[94]

* * *

Instead of holding firm to the belief 'that only if things work do they matter',[95] Rita and the other women had learned through their experiences the importance of working with what was broken, partial, uncertain, and what seemed doomed to end in destruction. They had learned this under the worst possible circumstances, of course. But, when they walked away from their experiences during the Second World War and the Holocaust, it was not only with their lives and more than a lifetime of trauma. They also walked away with sharply attuned instincts and ethics that contributed to their tenacity in their lives and work. They had been conditioned to stay with the trouble, at least partly through their experiences of surviving the Second World War and the Holocaust.

A Past Never Finished

Survival can happen in a moment, and it can take a lifetime. Those who survived Nazi persecution experienced both types of survival. Walking away from the Second World War and the Holocaust meant they had to find purpose and meaning that would sustain them in life. Halina Neujahr, Halina Czajkowska, Ellen Loeb, Ruth Pick, Rita Levi-Montalcini, and Eva Klein did not take this task lightly. Summoning all their strength and drawing on their experiences during the events meant to destroy them, they spent the rest of their lives running toward a future dedicated to STEM. Though they may have tried to forget the war and the Holocaust, as the quote by Eva and Georg that opened this chapter indicates, this was a past that was never finished. The tragedies and destruction led them to countries outside their homelands and into uncharted career paths at a time of immense change and discovery in science, medicine, and technology. Their unconventional training and experience in these areas during and after the war gave them unique abilities, sensibilities, and insights not shared by most of their peers. Faced in the aftermath with gender and social biases regarding women in STEM, these traits served them exceptionally well, not only in terms of personal survival but also career survival. None ever gave up, either on career or research, even in the face of doubts and dismissals. As a result, they made contributions to humanity that are evident even today. All six women are gone, but their pasts live on in these gifts.

Chapter Six

Feminist Trailblazers? Holocaust Heroines?

> Women often think in parallel paths and can do several things at the same time. Whether it is innate or socially acquired, I do not know. A good researcher can connect seemingly unrelated things. Women have excellent qualifications for that.
>
> Halina Neujahr, D.Tech, 1993[1]

The tendency to mythologise important and significant people in history has a reductive effect. Complicated lives are reduced to simple heroic narratives. Rarely can more than one aspect of a person's life fit into these reduced narratives. Nuance that humanises historical people is all but erased. Because heroes become less special when there are too many of them, heroic narratives also tend to perpetuate the idea that these individuals are unique and special rather than normal people who became notable. The result is that a handful of historical people can become representative examples of people in a particular context at the expense of many equally significant others. This book started out with these circumstances in mind, particularly the notion that women Holocaust survivors who were also STEM pioneers have rarely been recognised as both at the same time, and that their gender and the associated expectations of marriage and motherhood have further complicated how these women's Holocaust experiences, lives, and careers have been contextualised. Rita Levi-Montalcini at first appears as an exception – a woman recognised as both a Holocaust survivor and STEM pioneer whose unmarried and childless status in life has not appeared to be an issue. This appearance does not stand up to scrutiny, however. Although she has become in many ways a representative of both women in science and the Holocaust (thanks in no small part to her own autobiographical efforts), the two are often portrayed as two parts of a half rather than as elements of an integrated whole (even though she portrayed herself very much in the spirit of the latter). Moreover, the element of gender is at once obvious and obscured in how she has been portrayed by others.

In the same spirit as Halina Neujahr's above quote, it is important to acknowledge that women – no differently than men – are and can do many things at the same time, and these aspects of their identity and experience not only relate to one another but also overlap. Which begs the question: what were the women in this book and others like them? While it would be easy to place them in simple categories like 'feminist trailblazers' and 'Holocaust heroines', the complicated answers to the question are more interesting than summarising them in these terms, not least because once they survived the Holocaust and entered their careers in STEM, they were little different than any other woman working in these professions. Yet, at the same time, what distinguished them was their unconventional experience and 'training' during the Second World War and the Holocaust. This, combined with their gender, religion, and foreign backgrounds, created a complex combination of advantages and disadvantages in their lives and careers. To understand these women, it is therefore necessary to understand how they viewed themselves in terms of the social, political, and professional norms of the time, and where their careers and accomplishments fit in the context of the history of women in STEM. Did they view themselves as feminists? STEM pioneers? Were they conscious of how their experiences during the Second World War and the Holocaust impacted their careers?

Strong but Bendable

In 1997, Ruth Pick was interviewed by the Survivors of the Shoah Visual History Foundation (now known as the USC Shoah Foundation), a nonprofit organisation founded in 1994 by American filmmaker Steven Spielberg to record video interviews with survivors of the Holocaust (known in Hebrew as *Shoah*). As far as is known, this was the only time she told her life story for public consumption. Soon after arriving in the US, Ruth realised that few Jewish Americans wanted to hear about their European brethren's tragic experiences. 'We did not make it a big deal,' Ruth explained. 'If they wanted to know, we told. We didn't make a secret. We had nothing to hide. But people at that time didn't want to hear anymore. They heard so much.

It was too much.'[2] Perhaps because of this early experience with what is now sometimes referred to as 'Holocaust fatigue', Ruth appears to have maintained a quiet reserve about her experiences for most of her life apart from her 1997 video testimony.

Throughout her long career, few colleagues knew she was a Holocaust survivor. 'If they did,' wrote three male colleagues after her death, 'they did not hear it from her.'[3] One wonders if one of these authors – her colleague, cardiologist Earl N. Silber (1921–2009) – was aware of Ruth's Holocaust experience more than twenty years earlier, when he invoked the Holocaust during a group discussion on the risk factors and treatment of heart disease that was published in the *Chicago Tribune*. 'During World War II, inmates of concentration camps, exposed to the most intense, ongoing stress imaginable, suffered less, not more, coronary heart disease.' Ruth immediately responded, 'But that was group stress. We must distinguish that from purely personal stress.'[4] This exchange highlights the blurriness between Ruth's identity and experiences as a woman in medical science and a Holocaust survivor.

When and where Ruth came of age, women's careers in male-dominated fields like medicine were usually contingent on the acceptance of a patriarch and/or a patriarchal society. In her case, her father's influence was paramount. 'My father told me I have to learn something. He can't give me a big dowry, I'm not particularly beautiful, I won't get married. But he can give me an education, so go and learn something. He kind of wanted me to go to medical school,' Ruth explained. 'First, I was very interested in physics. I never got very excited about chemistry. He was very happy when I decided to go to medical school.'[5] As a woman in medical school, Ruth was in a minority. As a Jewish woman, that minority became even smaller due to the *Numerus Clausus* restrictions placed on Jews' enrolling in universities. Despite her father's prediction that she would never marry, she met her future husband while she was still a medical student. Since she completed medical school before she and Alfred married and she continued her career path afterwards, her husband clearly had no objection to her ambition to earn a medical degree. What she did after her marriage was another matter. As another Czech Holocaust survivor, Ruth Bondy (1923–2017), recalled,

> In normal times, the man was the breadwinner of the family. Married women, even if they had acquired a profession while still single, such as a teacher or secretary, became housewives or worked in the family business. Very few continued to work as self-employed or salaried professionals.[6]

However, there was nothing 'normal' about the times in which Ruth and Alfred Pick were living after Ruth completed medical school in April 1938 and the couple married in May 1938. Even before the annexation of the Sudetenland in September 1938 and the Nazi occupation of Prague in March 1939, political uncertainty and rising antisemitism made it difficult to establish a normal career path. Any objections Alfred might have had to Ruth pursuing a career in medicine would have been irrelevant under the circumstances. Ruth's work as a medical professional was a matter of survival from the moment she finished medical school. Working in the Jewish hospital in Prague delayed their deportation to Theresienstadt, where their medical work contributed to their survival. After the end of the war, medical work gave them shelter and food in dire circumstances in Prague. As immigrants in the US, both needed to work to repay the loans that enabled them to escape Communist Czechoslovakia and start life over in a new country. Throughout everything, it was never in doubt that Ruth needed to work. At the same time, she was in every scenario a woman in a man's world, both as a wife and a doctor, and would have become accustomed to the conditions and circumstances this engendered. While her response to Stamler early in her career at Michael Reese over the chick hearts demonstrates a steely determination to stand her ground against anyone – man or woman – who underestimated her, Ruth's career path was not fully in her hands. 'He [her husband, Alfred] did not allow me, and I don't know why – he did not tell me before he died – he didn't let me take an internship. So, I'm not licensed to practice medicine,' Ruth explained in 1997, with the qualification, 'But it did not make a difference in the laboratory.'[7] Eleven years earlier, she explained the situation slightly differently to a reporter for the *Chicago Tribune*:

> 'We discussed it,' she recalls, 'but it was so important to him that I not take an internship. I didn't argue long about it. It would have taken me longer to finish the training because I didn't have the experience he had. But to the last

day of his life, he never understood why it was important to me. Not that I would ever have left research. It's very satisfying work. But I wish sometimes I had had the internship.'[8]

All Ruth's work during and after the war – as a nurse in the Jewish hospital in Prague and as a physician in the Nazi camps and the psychiatric hospital in Prague – placed her on the path for a career as a medical practitioner, not a researcher. Alfred had also been prevented by the war from doing an internship in Prague, and his wartime and postwar medical work had followed a similar pattern to Ruth's. Yet, after they arrived in the US, only he took on a medical internship at Michael Reese. This was why Ruth made more money than Alfred did once she secured a salaried position; she was a technician, while he was a medical intern. Arguably, Ruth's strength lay partly in her flexibility. Her ability to adapt to the circumstances – to be both strong but bendable – enabled her to thrive in her life and career in the face of everything from Nazi persecution to her husband's inexplicable dictate.

By all accounts, Ruth had an extraordinary career in medical research, which may not have been the case had she practised medicine. In nearly every way, medical research was the more difficult path to take in Ruth's time. A woman physician who could not secure a position in a hospital could instead go into private practice. Medical research, on the other hand, meant working in the male-dominated domain of the laboratory where women were mainly technicians and laboratory assistants. In these environments, *all* women were subjected to gender discrimination and sexual harassment. It wasn't until federal legislation began to be introduced in the US in the mid-1960s – well into Ruth's career in Chicago – that things slowly began to change.[9] Tellingly, Ruth never mentions these factors when talking about her career during her 1997 interview with the USC Shoah Foundation. It is only in her anecdotes that they can be gleaned. Instead, she speaks proudly yet modestly of her contributions to medical research, especially to understanding of the relationship between oestrogen and heart disease, and the use of micro doses of aspirin for prevention of heart attacks. 'These are, in my own opinion, if I did anything that's worthwhile, these are the two things that I feel good about.'[10] It took others to more boldly assert her many other accomplishments, including as 'an outspoken, independent leader' and mentor.[11] She rose through the ranks during her fifty-year career at Michael Reese, eventually becoming Chief of the Atherosclerosis Laboratory in the Cardiovascular Institute. She was also an active member of the American Heart Association and served as President of the Chicago chapter of the American Heart Association.[12]

Ruth lived through but was not a product of the location and period in which her career flourished – the US in the 1950s to 1980s, which saw the rise and pinnacle of second-wave feminism. Her career path was determined in part by her husband, who did not 'allow' her to practise medicine. On the other hand, nothing prevented her from doing her job with passion and determination, overcoming gender discrimination in STEM in the process. Likewise, she defied the Nazis not by heroism but by a combination of luck and tenacity. Even if she did not share her Holocaust experience with others, it was a part of everything she did. As her three colleagues wrote in 2004, 'She had suffered more than any of us and seen atrocities that would have extinguished the spirit in most, and yet, she survived and contributed; she loved and taught; she inspired and led.'[13]

'The Queen of the NGF'

Rita's longtime colleague, Dr Luigi Aloe (1943–2025), referred to Rita just before her death as 'The Queen of the NGF'. He implied that this was because she built not only a scientific kingdom of sorts around NGF, but also a family. 'Rita Levi-Montalcini discovered the NGF by trying to investigate how the nervous system developed, but her study led to even more than that,' he wrote in a 2011 article.

> Building upon this type of approach, she, her collaborators, and a host of other researchers unearthed a whole class of biological mediators that are intimately involved in every single stage of a cell's or an organism's life from conception to death.[14]

While there's no doubt that Rita and her discovery of NGF were finally accepted by the scientific community, it is significant that both the struggle to achieve this acceptance and the acceptance itself were closely tied up with gendered notions of women's place in society. Throughout the period of struggle for recognition and acceptance, Rita was briefly mentioned in articles that appeared occasionally, often via national wire services, about women in science in America. The ambivalence towards women's role in science described by historian Margaret Rossiter is palpable in these articles, which reflect an uneasy preoccupation with women's ability to balance careers in science and technology with marriage, motherhood, and domesticity.[15] 'Probably the one question uppermost in the minds of young girls selecting a career is: "Will it interfere with marriage and my life as a woman?"' reads a 1961 article in which Rita was mentioned.[16] The extent to which this tension affected how women's scientific achievements were received is clear in a 1965 UPI news wire article about Rita. Reflecting her undeniable importance in the scientific community, Rita had just presented a lecture on Nerve Growth Factor at the prestigious Harvey Society in New York City, where she was interviewed by reporters. Though ostensibly about her lecture, the article focused primarily on Rita's thoughts on the barriers keeping women out of scientific careers in the US. 'Many bright girls are short-changed,' began the article. 'They're afraid to make extra efforts in their studies. Boys, they think, are afraid of bright girls. Dr. Rita Levi-Montalcini, professor of zoology at Washington University in St. Louis, said so in an interview.' The first half of the article continues in this vein, ultimately noting that Rita, who was about to celebrate her 56th birthday, 'feels that a career might be combined with marriage – for other people, not for herself. She has never married.' Only after these aspects are exhausted does the article elaborate on Rita's research and its importance, not to mention her demanding schedule of research, lecturing, and associated travel. The final sentence returns to what is clearly the main theme of the article. 'Looking at her schedule,' the journalist concludes, 'you see why combining marriage would be difficult for her – at this point.'[17] Had she been younger, the question of children would almost certainly have been discussed as well. Rather than focusing on the science, the primary concern was seemingly the marital and maternal status of women in science. Needless to say, the same was not true of male scientists.[18]

Rita would almost certainly have felt frustrated that she and not her science was the focus of attention. But that frustration would likely have been directed mainly at the neglect of her work rather than at the underlying gender inequalities that had been an ever-present theme in her life. 'I find it hard to remember when it was that I first became aware of the different roles that Father and Mother played in the running of the family, and of the enormous extent to which my father's will prevailed, unchallenged in decisions large and small concerning us,' Rita later recalled.[19] Her father's dominance of the family was not unusual for the time. What was unusual was that it was not based purely on blind adherence to social norms but rather on a reality that women struggled to have both a career and a family. Her father neither denied nor encouraged his daughter's abilities, partly due to the difficulties faced by his progressive sisters, 'who had – altogether unusually for the time – earned doctoral degrees in literature and in mathematics' and had subsequently 'found great difficulty both in going on with their studies and in reconciling them with marital and maternal obligations'. Even when Rita wrote these words in the 1980s, she observed that, 'Today, almost a century later, these same difficulties still exist, and the problem is far from being resolved.'[20] When, at the age of 20, Rita approached her father about her desire to study medicine, his reluctant consent was based only partly on embedded gender bias. 'He objected that it was a long and difficult course of study, unsuitable for a woman. Since I had finished school three years previously, it would not be easy to take it up again.'[21] On two of the three points, he was correct. Rita documented her struggles in these respects. As for the third concern – that medicine was not a suitable course of study for women – he was expressing a common belief that neither Rita nor any other woman could disprove, no matter how successful they were.[22] Fortunately, this was never Rita's intention.

By the fall of 1948, any idea of Rita leaving the US and returning to Italy was out of the question. Having overcome her doubts and insecurities, she set the course for her research that would ultimately lead to the discovery of NGF, with St Louis as her home base. No insurmountable obstacles stood in her way, either because she was Jewish or a woman. But while these factors did not prevent Rita from conducting the research that led to the discovery, gender bias undoubtedly contributed to the scepticism of the discovery among the scientific community, not to mention why for many years she was not cited in scientific literature related to NGF.[23] It was the lack of recognition caused by the gender bias in science that led Rita to briefly set aside NGF. 'At the end of the Sixties [sic] I just decided to go away and not do any more work on NGF. *If they want NGF*, I thought, *they can have it*.'[24] Earlier that decade, Rita began to divide her time and research efforts between the US and Italy. This move enabled her to distance herself somewhat from the scientific frustrations she felt in the US and spend more time with her twin sister Paola and their elderly mother in Italy. With the help of a US National Science Foundation grant, she established a research centre in Rome for the study of neurobiology, cell biology, and immunology.[25] During her break from NGF in the late 1960s, she turned to studying the nerve cells of cockroaches, through which she 'developed an entirely new field of research in neuroscience'. Despite her apparent success, she later claimed that she did this research 'just to get rid of this feeling of unhappiness at being completely ignored'. In 1972, Rita returned to research on NGF. 'I decided I could not give up what was my child, the NGF,' she explained in 1988, two years after winning the Nobel Prize in Physiology or Medicine for its discovery.[26]

Once she had achieved the ultimate recognition, Rita contributed to crafting her image around a gendered ideal as 'the mother' of NGF. In a 1992 scientific journal article, she even subtitled one section 'The NGF: Its Birth and Early Life History'.[27] After many years of being interviewed and featured in the international press, she would have been aware that if she did not craft her own image, others would. Instead of fighting the gendered notions about women in science, she embraced them, just as she had embraced the vicissitudes of the Fascist racial laws in Italy, the Second World War, and the Holocaust. By claiming a maternal role in relation to NGF, she was also following an accepted pattern of honouring male scientists with a paternal role as 'the father of' a field or discovery. When Luigi Aloe called her 'the queen of NGF', he was simply elevating her from the position she had given herself. Rita's rise to this level of recognition had been more difficult than the majority of her male peers, however, and not only due to her gender. Even in this respect, she was pragmatic. 'Yes, the obstacles were useful,' she responded in an uncharacteristically muted way when asked if the racial persecution she experienced contributed to her desire to excel.[28] More than useful, the obstacles lit a fire inside her. The racial laws connected her with her Jewish identity for the first time in her life. The death of her beloved governess inspired her to become a physician. When the Fascist racial laws prevented her from practising medicine, she turned to research. Her primitive laboratory encouraged rather than discouraged her. The anomalies that appeared under the lens of her microscope were wonders to be explored. Even amid her frustration at the lack of recognition from the scientific community, she told a journalist, 'I get a tremendous pleasure out of my work. My life has been all I could ask for.'[29]

'Can She Really Be a Scientist?'

'I'm not a feminist,' Halina Neujahr told a (male) reporter for a 1961 feature story in a Swedish women's magazine. But she clarified, 'I claim that women's opportunities in technology are just as great as men's!' She advised that women 'must take outdated prejudices under careful evaluation and careful consideration even at the risk of being considered a troublemaker with strange ideas'.[30] Halina's accomplishment of being the first woman in Sweden to earn a Doctor of Technology degree in chemistry in 1960 brought with it a significant amount of press coverage. Her professional career was followed in daily newspapers, and she was featured in several women's and general interest periodicals over the next few years. Women's magazines and newspaper sections were especially fond of her. Her reluctance to identify as a feminist but clear identification with women's equality to men was not unusual for the time, at least in Sweden. By the time Halina arrived in Sweden in 1945, Swedish women had secured many of the rights women in the US and UK were still fighting for in subsequent decades, including legalised birth control, liberalised/legalised abortion, and six months of job-protected maternity leave. These and other rights had been obtained not through radical action but because the goals of Sweden's leading political party, the Social Democrats, and the dominant women's rights groups had merged in the early decades of the twentieth century. The two groups joined forces 'to cast off the weight of industrialization and make *Moder Svea* (Mother Sweden) a harmonious country which would promote the ideas of democracy, humanitarianism and social equality'.[31] In seeking solutions for some of the major issues of the 1920s to 1960s, such as low population and birth rates, labour shortages and class conflict, the government found solutions in granting women greater rights and freedoms without their having to fight for these rights in the same way as women elsewhere have done. Of course, this did not mean that women in Sweden were (or still are) treated as the equals of men in every respect.

Nearly thirty-five years after denying being a feminist, Halina was part of a Swedish government inquiry into gender discrimination in Swedish higher education, which the resulting report described as having an 'archaic structure'.[32] With her career behind her – she was Professor Emerita at the time – and the benefit of hindsight, she summarised her experiences by saying, 'I did not feel bad as a woman or as a member of a minority until I started competing for positions.'[33] In addition to revealing that her career had not transformed her into a self-described feminist, it also highlights that she was aware of her 'difference', not just as a woman but also as a Jewish woman who was not born in Sweden. In the report of the inquiry, Halina's account began with the statement, 'my story is very different from the stories of others [interviewed for the inquiry]'.[34] She proceeded to recount her early life in Warsaw, her experiences during the Holocaust, and her time as a refugee and student in Sweden. This background was essential to the story of her career. To achieve what she did in her scientific career, she had to overcome far more than gender inequality. Had Halina's story only been about being the first woman in Sweden to earn a Doctor of Technology degree in chemistry, it's likely that her notoriety would not have lasted more than five minutes. But the inquisitive Swedish press quickly picked up on her full history when she made the headlines in 1960. Having kept her personal history to herself during her education, Halina 'suddenly became very famous. All these journalists rushed with cameras to take photographs of me,' she recalled with a laugh.

> Then everyone became curious about my background. And then I was, of course, forced to tell about my background. I couldn't fabricate or tell lies. So I started to tell – in a restrictive manner – but I started to tell, and they started to blow it up and make a story like journalists do.[35]

Halina became notable not only as a woman who achieved something extraordinary in a male-dominated field but also as a woman who survived Nazi persecution *and* achieved something extraordinary in a male-dominated field. It was almost a best-case scenario in terms of acknowledging different but intersecting aspects of her life and career to that point. Almost because of her gender. 'From the concentration camp to the university and the doctorate in technology,' read the headline of a feature article about Halina in the Swedish women's periodical *Vi Kvinnor* (We Women) in 1961. A promising start until the first paragraph, written by a male journalist, cited here in full:

> Can she really be a scientist, this little, highly feminine creature in a soft home dress? In that case, one must quickly revise one's perception of female scientists, whom one has always thought to be highly masculine, but Halina is femininity personified. She laughs happily, low and musically, her green eyes sweep over the room and the visitor with an intelligent and curious feminine interest, she speaks easily and fluently. An extremely weak piquant accent betrays her foreign origin, for she is a foreign bird, Halina Neujahr, a Polish Jewess, whom the angry waves of the Second World War washed up on our shores, one of the few who survived the systematic extermination of the Jews by Nazism and who, with her talent and indomitable will as her only capital, has created a top position for herself in the world of technology, an area that, at least here, is reserved for men.[36]

As for Rita, it was not the science or the scientist but rather the anomaly of her status as an unmarried woman that was of primary interest to journalists and, by extension, the public. For Halina, it was as a survivor and 'foreign bird' that she became a media sensation. Science only came along for the ride, so to speak, because it made them seem even more unusual and singular. We begin to see why Ruth kept her Holocaust experience to herself and why Halina only reluctantly began sharing it when approached by curious journalists. She and the other women simply could not be conceived or understood as everything at once: women, survivors, scientists. Of course, in Halina's time, this would have been simply part of the human condition, the normative way of being, and a standard way of addressing differences. These realities would not have been lost on Halina, even if she may have disliked the portrayal, something we are never likely to know. We do know, however, that Halina understood and even emphasised that a person – *women* to be precise – can be and do several things at the same time, as the quote that starts this chapter demonstrates. And this is exactly what Halina did.

Once her Holocaust experience was out in the open, Halina first accepted and later embraced it as a part of how her career and life were narrated and lived. In interviews conducted over the next several decades, Halina recognised that her Holocaust experiences were part of her career as a researcher. In 1971, she gave a relatively complete history of her Holocaust experiences to a major Swedish daily newspaper, including her clandestine medical studies in the Warsaw Ghetto and the terrible incident when the concentration camp guard pushed her head into raw sewage. Based on the interview, the headline read, 'She began her research career in the ghetto.'[37] Though not completely accurate, the headline nonetheless captures the reality that Halina's Holocaust experiences were an embodied part of her life and career. This acknowledgement would not have been possible without Halina's candid retelling of her experiences, which she described on multiple occasions as 'humiliating'. To bring together her experiences and emphasise their wholeness within her, she revealed both her vulnerability and her strength. She understood implicitly that vulnerability and agency are not mutually exclusive, and that agency can even at times be born of vulnerability.[38]

Through the 1970s and 1980s, Halina continued to be recognised in the Swedish press as both a scientist and a Holocaust survivor, sometimes simultaneously. This began to change in the late 1980s and early 1990s, when an increase in antisemitic sentiment and violence in Sweden transformed her again into a moral witness, whose story of suffering and survival is shared for the benefit of others.[39] 'She survived five concentration camps,' read the headline of a 1992 feature story on Halina in the Swedish national daily newspaper *Expressen*. Not only had her career disappeared from the headline, it was also treated cursorily in the article. In this case, Halina probably didn't mind, since her purpose for speaking out now was to fight antisemitism and Holocaust denial, not to link together her Holocaust experience with her career. 'Now that people are showing up who want to deny that this happened, several of us survivors have become so upset and offended that we felt we had to do something,' she explained.[40]

But just because the narrative changed does not mean that she did. She continued to do several things at once. While thriving in her career, she simultaneously fought for her professorship at KTH, struggled with long-term physical and mental health issues, applied for reparations from the German government, took university courses on the history of ideas and religion at Stockholm University, and dedicated herself to remembering and commemorating the Holocaust through public speaking, writing, and helping to establish a Holocaust memorial in Stockholm. Even after her retirement in 1991, she continued to work in the laboratory provided by KTH and increased her Holocaust activism until late in her life. In a 1987 newspaper in a Swedish newspaper about a talk that she gave to students, Halina answered a student's question about how she managed to study and catch up on her education as a refugee by saying, 'You either lose your mind, kill yourself or decide to deal with it. My studies and my work have helped me. It was the only thing I had the strength to do when I came to Sweden.'[41] Halina was always more than one thing, always walking parallel paths, in her life and career.

Life After Purgatory

Halina Czajkowska had the dubious privilege of living to see herself both left out of and written back into history. Nearly half a century passed between these events, and Halina filled this time with many joys and accomplishments. Among the first was arranging for her parents, who were still in Sweden, to join her in Ontario, Canada. 'Still with only rudimentary English and living in a small rented room within a private home, she sought a bank loan of $500 to secure their paperwork and voyage,' explains the editor of her memoirs, Jacalyn Duffin.[42] Duffin suggests that Dr J. B. Collip, the head of Collip Laboratory at the University of Western Ontario (UWO), where Halina worked, was the mystery individual who guaranteed the loan for Halina, a recent immigrant. The newly reunited family initially lived in a rented apartment while Halina worked at the Collip Laboratory. More joy came out of her time at UWO when she met and ultimately married James Russell Robinson, who was a doctoral candidate in chemistry at UWO when he asked for an introduction to Halina, a request fulfilled by Collip. Halina and Russell were married on 1 August 1953. Halina's published memoir – titled *Heaven, Hell, and Purgatory* – ends before these events, which Duffin picks up in a brief epilogue. In one of the final paragraphs of the memoir, which she completed in January 2002, Halina writes that when she and her mother were reunited with her father in Stockholm in December 1948, 'our purgatory was nearly over'.[43] It raises the question of what she would have called the period that came after purgatory, which began after the liberation of Bergen-Belsen in April 1945. Heaven had been her childhood before the Second World War, which was of course Hell. It is tempting to romanticise the early 1950s then as another Heaven, a Paradise of sorts when she was reunited permanently with her parents, made an important contribution to medical science, and married her life partner. The reality would have been far more complicated. The struggle to find her place professionally, to learn English, to support her parents, to date (something the circumstances of her youth and early adulthood had made all but impossible), would have entailed elements of heaven, hell, and purgatory.

Halina left the Collip Laboratory in the mid-1950s to take the position of 'Chemist-in-Charge' of the Pediatric Research Laboratory at the War Memorial Children's Hospital in London, Ontario (now the Children's Hospital at London Health Sciences Centre), where she remained until her retirement in 1976 or 1977.[44] 'She made important contributions in the analysis of blood sugars and electrolytes, and in the development of micro-chemical analytical methods suitable for the few drops of blood available from newborn infants,' explains Duffin, who is both a haematologist and a medical historian. 'In an unpublished study, she discovered that the Respiratory Distress Syndrome ("Hyaline Membrane Disease") in premature infants is due to immaturity of lung tissue and not to the inhalation of amniotic fluid, as was previously thought.'[45] With a full and meaningful scientific career behind her, Halina turned to a new career in fine art, first attending and then teaching art classes as well as working as a critic and tour guide, and of course creating her own fine art. From the 1970s onward, her watercolours and woodcut prints were exhibited many times in Ontario.[46] During this productive period of her 'retirement', Halina came to terms with her past accomplishments.

Although she had been unable to speak up during Dr Robert Noble's speech about the discovery of vinblastine, in which he could not answer the question, 'Why ever did you look at the white [blood] cells in the first place?'[47] Halina was eventually persuaded to write and publish a brief article published in 1991 that explained her role in this part of medical history.[48] Responding to Noble's journal article published in the same scientific journal the previous year, Halina wrote, 'I wish to place on record one more anecdote which is crucial to a correct and complete account of the role of chance in these events.'[49] She explained in the article her reasons for conducting the white blood cell counts, referring especially to her experience at KTH in Sweden. As a result, she wrote, 'I was the first to observe, and to report to Dr. Noble (in 1953), that blood cell counts were profoundly affected by the *Vinca* treatments.' With justifiable confidence, she pinpointed the significance of her contribution: 'The fortuitous observation of the WBC [white blood cell] depression gave new impetus to "the Jamaica plant project" [as the research had been named] and paved the way for the 1958 discovery of vinblastine in Canada.'[50] Her approach was both assertive and tactful. She made no accusations of gender discrimination or the invisibility of women working as laboratory technicians, a phenomenon historian Julie Des Jardins describes thus: 'Nobel Prizes went to the men who theorized the ideas women conscientiously confirmed at the bench.'[51] Halina neither accuses Noble and Beer of deliberately ignoring her, nor does she take credit for the discovery of vinblastine. Rather than calling out the narrative Noble established early on that replaced Halina's findings with the element of 'chance', she embraced it.[52]

> Throughout the *Vinca* story (1958–1990), 'the role of chance observations' in the discovery of vinblastine as a chemotherapeutic agent against cancer has been repeatedly highlighted. It is time, for historical accuracy, to supply this missing detail on what was perhaps the most crucial chance observation of them all [i.e., her own].[53]

Despite Halina's intervention in the history of science and medicine, she was not acknowledged when Noble and Dr Charles Beer were inducted into the Canadian Medical Hall of Fame in 1997 for their joint discovery of vinblastine. The fullness of Halina's story came about thanks to Jacalyn Duffin, who interviewed Halina for her comprehensive corrective history, published first in 2000 in the scientific journal, *Canadian Bulletin of Medical*

History, for which Duffin was awarded the J. Worth Estes Prize of the American Association of the History of Medicine in 2002. An expanded version of the article was then published in 2002 in the journal *Pharmacy in History*.[54] This important work also brought Halina's experiences during the Second World War and her time as a refugee in Sweden into focus for the first time. With Duffin's encouragement, Halina began writing her memoir in 2000. Thanks again to Duffin's efforts, the memoir was finally published in 2020. Halina thus lived to see both her scientific legacy secured and her wartime and postwar experiences brought to light, elements of her life this book contributes to integrating. Today, Halina is recognised on the Canadian Medical Hall of Fame website for her contribution to the discovery of vinblastine. Although she does not have her own profile page, as Noble and Beer do, each man's profile includes a mention of Halina. The passage on Noble's page reads:

> 1952. Noble's research focused on potential anti-diabetic properties of the periwinkle plant extract, however, he confirmed that it had little effect on the blood sugar levels of experimental rats. Laboratory technologist Halina Czajkowska observed, however, that the extract destroyed the rats' white blood cells and bone marrow. This finding raised the possibility that the extract could be used to treat leukemia, a cancer characterized by high counts of abnormal white blood cells.[55]

At no point in her life story does Halina claim to have been an advocate of women's rights, women's equality in STEM, or the Second World War and the Holocaust. Her actions speak volumes, on the other hand. When the German occupiers stripped Poles of their right to an education, Halina fought for hers, risking her life in the process. There was never a question for Halina that an education was her right as much as anyone else's; a universal right, not a gendered one. After the war, she continued to pursue her education without consideration for Swedish gender norms about women in engineering. In this sense, being an 'outsider' in Sweden was a benefit – as it was with Halina Neujahr. Unconcerned by the Swedish conception of engineering being a 'boys' club', they followed their inclinations and aptitudes rather than redirecting themselves into more ostensibly 'feminine' areas of study. When Halina Czajkowska left Sweden for Canada, she kept to her chosen path in seeking work as a chemical engineer. When this turned out to be impossible, she – like Ruth – remained flexible but strong by getting employment where she could but also striving to maintain her most valuable skills. Overlooked for her contributions to the discovery of vinblastine, she claimed her place with dignity and style. All these actions led to her not only becoming a STEM pioneer but also being recognised for it during her lifetime.

'How Did You Manage?'

In 2011, during a joint interview of Eva and Georg Klein for the American Association for Cancer Research, Eva referred to the tensions between her various roles in life.

> In fact, it is quite interesting, whenever I visit a laboratory of our colleagues and I would like to talk about science, and when it is a lady I'm talking to, the first question is not about my work. They look at me and say that I heard that you have three children and you work with your husband, how did you manage? So I can understand still that it was not… it's quite good for me to know that it's a general attitude and general understanding that my situation must not have been so easy all the times.[56]

The question of how Eva 'managed' to be a scientist, wife, and mother is often asked of her in interviews. Eva never flinched when responding to what we now might consider a sensitive question. Both Eva and Georg were also strikingly candid about inequalities in the marriage. 'She had the main burden, and he had the bad conscience,' according to one interview-based profile of the couple. Not only was Georg not an active parent, but he also expected Eva to not be one either. Although they had a nanny, this was not a luxury but a necessity since Georg explicitly did not want Eva to be a housewife, only a scientist. Eva thus existed in a sort of gender paradox, both supported and not supported by her husband, who viewed her mainly as a scientist and wife at the expense of her role as a mother. Later in life, she openly acknowledged that her children resented their parents' lack of involvement in their childhoods. However, Eva justified this by explaining her commitment to science and to Georg.[57] It perhaps goes without saying that Georg was never asked how *he* 'managed' to be a scientist, husband, and father. By his own admission, on the other hand, he was considered a feminist.

> Our lab was always non-hierarchical, non-ethnicist, non-racist, non-sexist. One of my female doctoral students said 25 years ago that I should be an honorary president of a feminist emancipation organization because I had never shown the slightest tendency for sex discrimination.[58]

By 'our lab', he was referring to his and Eva's laboratory at KI, where presumably they, and not just he, set the tone. Yet the egalitarianism in the lab may only have reached so far. Eva recalled that when their laboratory which was established at KI grew beyond just her and Georg working in partnership, the dynamic became strained, at least for her.

> I had a kind of a situation which was not so easy, because [Georg] was the boss. And I had to mark that whatever I say is not his opinion. And also one of us… I couldn't think about my reputation or whatever… I had to see that the institute, that the show should go [on]. So many times, I stepped down… […] The attitude was that, I mean, his things are more important.[59]

In other words, Eva felt she had to – and did – take a backseat to Georg in the laboratory. Significantly, she attributed her willingness to let Georg lead in every way to her wartime experiences. 'The education and the attitude of women in the generation when I was born […] the man was much more important,' she explained in her 1997 USC Shoah Foundation interview. 'The show was in the hands of the man, though I would never admit that in my youth. But the war changed this in me.' By her own admission, her experiences of hiding during the war made her seek an untroubled, 'well-organized' life, something she continued to crave even later in life. Letting Georg make the decisions and take the leading role – or at least not actively resisting him in these respects – ensured peace and stability in her life. At least in her scientific work, Eva managed to do things her own way. 'Our way of working is very different,' she explained of herself and Georg.[60]

> We also have very different scientific styles; for example, I write very slowly and he does it very fast. He does not like to get involved with details; he is irritated by them; I like to ponder over small details as well because you never know where the solution of a problem comes from. Actually we are so different, we do not think the same way, so our joint work should be ideal.[61]

Had Eva left decisions about her research to Georg, she might never have discovered NK cells. 'She stubbornly wanted to look at what Georg called "the background," in which no one else was interested. And, as it turned out, she was right.'[62] Despite their differences, or perhaps because of them, Eva considered that she and Georg were partners in their research. Although this was true, they were not always recognised as equals. In 1957, the Department of Tumour Biology at KI was established with Georg, who was granted the title of Professor, as its first chair. It took another twelve years before Eva was also given the title of Professor in 1969. Although she and Georg continued to collaborate, Eva increasingly followed her own research paths, gaining funding and hiring and supervising researchers and graduate students independently of Georg. 'Eva has always attracted a large group of research fellows for whom she has been an inspiring mentor,' wrote a Hungarian colleague, who first met Eva at medical school in Budapest.[63] In 1993, she formally retired and became Professor Emerita at KI, though she continued to work for more than a decade. Recognition of Eva's personal significance to cancer immunology eventually came to her. During her long career, she became a member of many professional organisations and, like Georg, was a member of the Nobel Assembly at KI, which selects Nobel Prize laureates in Physiology or Medicine, and of the prestigious Royal Swedish Academy of Sciences. She was also recognised in her native Hungary, becoming a member of the Hungarian Academy of Sciences and the Hungarian Immunological Society.

Eva and Georg had vowed soon after arriving in Sweden that they would put the painful past behind them, and their Holocaust experiences were not publicly evident during the early stages of their careers. But they quickly realised it was not possible simply to forget. 'We could not suppress the memory of our murdered relatives, nor the two-thirds of our Jewish classmates and friends who never came back from the slave labor camps.'[64] Together and individually, they wrote about their Holocaust experiences, including how they were intertwined with their career paths as scientists. In 2000, Eva and Georg co-authored a review of the poetry by Polish American theoretical chemist and Nobel Laureate in chemistry Roald Hoffman (1937–). The review was titled 'The scientist as poet'. Like the Kleins, Hoffman also survived the Holocaust. As a young boy during the Nazi occupation of Poland, he, his mother, and other family members were hidden in an attic of a schoolhouse for eighteen months by a Ukrainian teacher, an act that ultimately saved them. Hoffman's father was murdered by the Nazis. What struck Eva and Georg about his poetry was that it 'has three dominating, inseparably interwoven themes: his childhood in the attic, chemistry, and love', they wrote.

> Chemistry has come to provide the scaffold, the structure that transforms the mood into poetry. Yes, chemistry of all things, no mistake about it. It provides a solid, rational basis on which the pains of the past and the present can rest and be transformed into words.[65]

Eva, who had spent part of her time hiding in a histology laboratory and found her purpose in cancer research alongside her husband, might have been writing about herself. Science was the structure on which much of her life and love had rested. It played a role in her survival during the Holocaust, finding her life partner, escaping Communist Hungary, and giving her life purpose and meaning. No structure can support everything, however. Eva was aware that she did not and could not manage everything. As a woman in science, her choices were difficult and society's expectations of her were (and still are) unrealistic. Though not superhuman, she was an exceptional person whose contributions to science and Holocaust experiences were closely intertwined. Twenty-five years after Eva and Georg lauded Hoffman as the scientist as poet whose Holocaust experiences deeply affected his life, Eva's colleagues recognised her in much the same way.

> She wrote short essays, and after retirement, while continuing her research as an emeritus professor, she translated poems from Hungarian to Swedish. In the essay 'Eva Climbs over the Grave' (in 'Dear Eva, Dear Georg', Karolinska Institutet, 2023), she tells the compelling story of her early life in

Budapest. She describes a privileged and creative environment, her dreams of athletics, literature, and art… and the coming of Nazism, racial laws, the escape from deportation and brave rescue of a beloved brother, and the loss of friends and family. She would realize her dreams in a new homeland. Life turned out differently from young Eva's dreams, but art and science were always intertwined.[66]

A Woman With It All?

When Ellen moved to McKinney, Texas to begin her internship at the VA Hospital, her mother, Dina, went with her. Later, the two women returned to Dallas when Ellen finished her internship and took up her residency at the VA Hospital in Dallas. According to Ellen's sister Trudy, Ellen and Dina had an active social life. 'She and my mother had a wonderful time,' Trudy recalled. 'They had made friends with lots of young doctors and their wives, and my mother was their adopted grandmother. They had a good life.' In general, Ellen made many friends, Trudy recalled, and was well respected professionally.[67] Even years after she left McKinney, she remained well respected there, with one local newspaper reporting in 1964 when Ellen was a guest speaker at local women's literary social club meeting that it was 'especially interesting since Dr. Loeb was once associated with the VA Hospital here and made many friends while in our midst'.[68] After joining the Wadley Institute in 1956, Ellen's public profile began to grow in the Dallas area due to her clinical research on leukaemia and treatment of patients with the disease, especially for her development and use of new treatments for leukaemia such as L-Asparaginase. She was a frequent speaker at events, raising awareness of and helping raise money for the Wadley Institute. Attitudes about female physicians were slowly changing in the 1950s and 1960s, and the newspaper articles about Ellen demonstrate that she was taken seriously as a medical professional.

Professional respect and journalistic courtesy aside, the situation for professional women was still quite grim in Texas and elsewhere in the United States throughout Ellen's career. A prominent male physician in the US claimed as early as 1958 that 'the young woman of today who aspires to become a physician will be given every opportunity to achieve her ambition on a plane of equality with men'. The reality, however, was that American women were far from being on an equal footing with men.[69] Pioneering Texas journalist Vivien Castleberry (1922–2017), a contemporary of Ellen's, wrote of Texas and Dallas before the establishment of national equal rights legislation in the mid-1960s to 1970s:

> ...to be born female in Texas was to be legally and socially bound, first to your father and then to your husband. By law, three classes of people were incompetent to give legal consent in those days – idiots, imbeciles, and women. The only way a woman could manage her own affairs, including property she inherited, was to have her incompetent status removed by law, to be widowed, or to be a mature single woman. Even then, legally, a single woman or a widow needed the cosponsoring signature of her father, a brother, or an adult son when she was buying, selling, or exchanging property of any kind. Dallas looked male and felt male. Banks, law firms, city offices, publishing companies, retail stores, churches, schools – everything – had a man at the top who was supported by women in his life – his wife, his mother, and his private secretary.[70]

This was Ellen's reality, even as a highly educated professional woman, a status that presented its own issues. As Castleberry recalled:

> In those dawning days of the women's movement, only three kinds of women worked: single women who were biding their time until Mr. Right came along and rescued them; women forced by economic necessity to earn a paycheck, who often resented it and carped about the injustices of life; and the rest of us, the handful who chose careers instead of, or in addition to, husbands and children. We had come of age during World War II, had savored independence, and had refused to go back to the kitchen again even when the popular press was admonishing us to find total fulfillment within the four walls of our homes. We were out of step with our times and scapegoats for all of the world's ills- divorce, depression, and delinquency.[71]

Ellen's chosen path in life had put her in the third category, and although there is no record of her proclaiming herself a feminist, her long association with the Business and Professional Women's Club (B&PW) aligns her with Dallas women's fight for equal rights. Amidst a plethora of women's social clubs, the Business and Professional Women's Club of Dallas was founded in 1916 – four years before the ratification of the Nineteenth Amendment to the US Constitution in 1920, which prohibited denial of US citizens' right to vote on the basis of gender – to promote the interests of professional women and, more generally, equal rights for all women. The First World War caused the organisation to lapse until 1919, when the National Federation of Business and Professional Women was established, with the Dallas club becoming one of many local chapters around the country. Due to its origins in Dallas, many Texas women served in leadership roles in the national organisation. Among the most prominent members of the Dallas club during Ellen's time was Judge Sarah T. Hughes (1896–1985), Texas's first woman state and federal judge, and the only woman to administer the oath of office to a US President – Lyndon Baynes Johnson, on 22 November 1963, aboard Air Force One at Love Field in Dallas, following the assassination of President John F. Kennedy.[72] B&PW and its leaders and members have historically been staunch advocates of the Equal Rights Amendment (ERA), an amendment to the US Constitution first proposed in 1923 to prohibit sex discrimination that has yet to be ratified. Ellen's connection with B&PW appears to have begun in 1953, when she addressed members of the McKinney club at an event. Between 1962 and 1977, she was a member of the Dallas club. During this period, she, like other members of B&PW, achieved her professional success against significant odds and inequalities and was dedicated to improving conditions for other women.

When Ellen wrote her letter to Trudy soon after the end of the war, detailing her Holocaust experiences, she explained that she would 'recall it [her experiences] one more time so I won't forget'.[73] According to Trudy, Ellen told her that 'neither she nor my mother wanted their story told'.[74] It is unlikely, then, that Ellen spoke publicly about her Holocaust experiences – for example, during her professional talks. Yet she was clearly open about her Holocaust experiences with her patients and journalists. As early as late 1958, one of Ellen's patients at Wadley referred to Ellen's Holocaust experiences, writing to a local newspaper in Texas of Ellen 'having survived three tortuous years in an [sic] European concentration camp'.[75] In 1961, Dallas-based United Press International (UPI) journalist Preston McGraw interviewed Ellen for an article that appeared in national newspapers with some variation of the headline, 'MD Escapes Gas Chambers: In Gratitude She Has Devoted Her Life to Medical Research.'[76] Though she may never have related her full story again after her letter to Trudy, she did not hesitate to link her Holocaust experiences with her medical career and practice. As she told McGraw in 1961, she had learned during the Holocaust to never give up, and this was a lesson she passed on to her patients.

Around the time the UPI article on Ellen appeared in national newspapers, psychotherapist Mortimer 'Mort' Katz moved to Dallas from New York. Mort was born in New York in 1925 to Jewish immigrants. Both his parents had emigrated to New York as children with their respective families; his father, Morris Katz, from Romania, and his mother Sophie Katz, née Guttman, from Lithuania. He served during the Second World War in the Philippines in the medical corps and earned a Master of Science in Social Work from Columbia University in New York, with a specialisation in psychiatric social work. He moved to Dallas in 1961, where he worked as a psychiatric social worker at the Dallas State Mental Health Clinic.[77] Whether he and Ellen met through social, religious, or professional circles is unclear. Quite possibly, they came to know each other through one or more of these. Regardless, in early June 1964, Dina Loeb announced her daughter Ellen's engagement to Mort Katz. On 18 July 1964, the couple married in a small Jewish ceremony at Ellen's home in Dallas. She was 44 years old, he was 39. Trudy was matron of honour, and her husband Rudy gave the bride away. After a honeymoon trip in Colorado and Wyoming, they settled into Ellen's home.[78] After marriage, Ellen continued to be known professionally as Dr Ellen Loeb or sometimes Dr Ellen Loeb Katz. The marriage appeared to be both happy and mutually supportive. In 1968, Mort left his job at the Dallas State Mental Health Clinic and started a private practice for individual, group, and family counselling. Ellen and the marriage were a source of inspiration to Mort and even his clients. In a 1972 newspaper profile of one of Mort's weekly women's group therapy sessions, one of his clients is reported as saying, 'I think that his wife, Dr. Ellen Loeb Katz, cancer specialist, must be a compassionate woman who has helped his understanding of women.'[79] For his part, Mort supported Ellen's career and the causes she held most dear. In 1970, both Ellen and Mort were part of the newly established Leukemia Association of

North Central Texas, which raised money for leukaemia patients and their families in Texas and for research.[80]

In 1969, Ellen became Chief of Medicine at the Granville C. Morton Cancer and Research Hospital, a division of the Wadley Institute of Molecular Medicine (as it was then known). At age 49, she was at the top of her profession and happily married. She loved her work and was surrounded by a supportive family. She seemed to 'have it all'. She had achieved work–life balance and success in a man's world and male-dominated profession. Perhaps she did. What is certain is that she never fought only for herself, a characteristic that can be traced back at least to her time as a nurse in the Nazi camps, when she fought for her life and the lives of all those she encountered, no matter how hopeless their situation. Ellen's career success was based on her dedication to her patients and finding cures for leukaemia and other diseases. Her advocacy for women was for the benefit of all women, to ensure that those who came later did not have to confront the same challenges. Having survived the Holocaust, she used her experiences to give hope and healing to others. It's hard to say if she truly 'had it all' but it is certain that she gave her all to others.

Parallel Paths

To paraphrase Halina Neujahr, women exist in parallel paths and are several things at the same time. The women in this book were women, survivors, and scientists. A few were married, one had children. Some considered themselves feminists, others did not. Their professional lives and accomplishments were heterogeneous. None identified as a trailblazer. Each experienced the Second World War and the Holocaust differently. None saw herself as a heroine. They defy categorisation and easy description because they were – as we all are – complex individuals who struggled at times to know themselves. And yet each one of them understood the value of her experiences during the Second World War and the Holocaust and recognised her important contributions to STEM. Ruth was proud of her work long after it was forgotten. Rita understood long before others that she was the 'mother' of NGF. Halina Czajkowska was the first to set the record straight on her contribution to science. Eva asserted herself as a scientist and made her own mark independent of her husband. Ellen reached the pinnacle of her profession not through ambition but through compassion. Halina Neujahr overcame one of the most humiliating experiences of her life to make her contributions to science, finding beauty in the process. This is how the women can be remembered in their own unique ways as women STEM pioneers who survived the Holocaust.

Epilogue

May Their Memory Be a Blessing

Zichronam L'Bracha.
Hebrew saying meaning 'May their memory be a blessing'

Ellen Loeb Katz, MD (1920–1980)

On 17 April 1980, eleven years after she became Chief of Medicine at the Granville C. Morton Cancer and Research Hospital in Dallas, Texas, Ellen died there of ovarian cancer at the age of 60.[1] Her death prompted her husband, Mort Katz, to write a booklet about dealing with grief dedicated 'to the loving memory of Ellen Loeb-Katz, M.D.'.[2] Dina Loeb, née Mayer, outlived her daughter by nearly a decade, dying at the age of 94 on 17 January 1989. Gertrude 'Trudy' Shakno, née Loeb, Ellen's sister, died on 25 August 2012, at the age of 98. Trudy's husband Rudy Shakno died in 1986. Mort Katz died in 2003, age 78. The couple had no children together. Ellen was the first to be buried at the Sparkman Hillcrest Cemetery in Dallas, where she was later joined by Dina, Trudy and Rudy, and Mort.

In October 1969, Ellen returned to Wuppertal, Germany, where she was born and lived until she and her family escaped to the Netherlands in the 1930s. It was work and not nostalgia that brought her there, but she could not avoid the latter. She was attending the Symposium on Experimental and Clinical Effects of L-Asparaginase, held at Bayer Pharmaceutical in Wuppertal, where she gave a lecture titled 'Treatment of Acute Leukemia with L-Asparaginase', based on clinical investigation and treatment she and her colleagues conducted at Wadley.[3]

> As she addressed the distinguished audience, Dr. Loeb looked across from the lectern through the large windows facing her. After her lecture, she informed them that the house directly in her line of vision was that in which her father

> had practiced medicine before he had been forced to give up all he had and flee because he was a Jew.[4]

The effect this had on the audience must have been as powerful as it was unexpected. In a minute or possibly less, Ellen communicated both the tragedy and devastation of the Nazi racial policies and her own incredible resilience. But then, this was something that she did throughout her medical career – she shared her experiences with her patients and others in small, healing doses, as she might have done with medical treatment.

Ellen died before the national and international campaigns of the 1980s and 1990s to collect audio and video testimonies from Holocaust survivors. Fortunately, Trudy Shakno saved Ellen's correspondence, including her comprehensive account of her experiences written in 1945 and 1946, as well as many documents and personal effects, including Ellen's concentration camp uniform, all of which are now held by the Dallas Holocaust and Human Rights Museum.

Ruth Pick, MD (1913–2003)

Ruth died of cardiovascular disease at the age of 89 on 19 July 2003, in Chicago. As was the case with Ellen, she could not escape the fate of the disease she spent decades researching. She was buried quietly with no funeral service. She had outlived Alfred Pick, who died in January 1982, by more than twenty-one years. The couple had become US citizens in 1954.

Right up to her final year of life, Ruth continued to visit her laboratory at Michael Reese Hospital every day.[5] She kept mentally sharp by doing the *New York Times* crossword puzzle and watching the long-running American quiz show *Jeopardy!* As one colleague remarked of her daily viewing of the popular show, she 'answered questions before the contestants would'.[6] Ruth and Alfred did not have any children, and they appear to have lived a quiet private life in Chicago. Although there is little evidence of their life together, journalist Lilian Pompian's description of Ruth and Alfred in the *Chicago Tribune* in 1959 paints a sweet picture of how she perceived their domestic life. It reads in part:

> A shy but charming couple, they live in a pleasant north side apartment containing such intriguing objects, salvaged from their travels, as a hand carved Chinese camphor chest and a Turkish waterpipe inlaid with mother of pearl. […] Dr. Ruth is wistful about being a 'good housekeeper,' an accomplishment which she feels has eluded her. But the orderliness of the apartment and the fragrance of roasting duck wafted from the kitchen belie her modesty. The intricacies of broiling a steak, however, confounded the petite scientist. 'In Europe, we do not have broilers. I have not even a word in my vocabulary for it.' And she cannot bake 'a real cake.' 'It collapses,' she says tragically.[7]

Besides Alfred, Ruth's only immediate family member to have survived the Holocaust was her younger sister Gertrude Lichtenstein, who died in 1988. Like Ruth, she settled in the US, where she remarried (her first husband having perished in the Holocaust) and had a daughter. Ruth's father, Arthur Holub, had died of a heart attack in 1937, but her mother, Paula Holub, was murdered in Auschwitz, probably on arrival in October 1944. Alfred's parents, Richard and Hedwig Pick, both perished in the Treblinka extermination camp.[8]

Halina Neujahr, D.Tech (1924–2006)

Halina Neujahr died in Stockholm, Sweden, at the age of 81 on 24 July 2006, sixty-one years after she arrived in Sweden. She did not marry or have children. The only other member of her immediate family to survive the Holocaust was her eldest sister, Rachela Langler née Neujahr (b. 1918), who was living in England when the Second World War began. After the war, Rachela moved to Sweden, and she died in Stockholm in 1999 at the age of 80.

In addition to Halina's scientific legacy, she contributed a great deal to Holocaust remembrance and education in Sweden through her public speaking and writing. She was instrumental in the effort to gather the support and funds necessary to create a Holocaust memorial in Stockholm, which was realised in 1998 under the auspices of *Föreningen Förintelsens Överlevande i Sverige* (the Association of Holocaust Survivors in Sweden), for which she served as a board member. Covering a wall within the courtyard of Stockholm's Great Synagogue, the monument lists the names of approximately 8,500 European Jews – relatives of Jews in Sweden – who were murdered during the Holocaust, including Halina's family.[9] In a 1995 book edited by Halina's KTH colleague and friend, Roman Wroblewski, whose parents were Holocaust survivors, Halina contributed an essay describing her and her family's experiences during the Holocaust. Her final paragraph emphasises the need to remember the Holocaust and restore the victims' identities, as the monument she helped to erect aims to do:

> Nazism's total war against the Jews was carried out systematically in different stages. Initially, all Jews were demonized. Then they were stripped of, in turn and order: civil rights, property, freedom, health and individual identity. To constantly review this process… to constantly analyze the psychological and social mechanisms, all of which led to the macabre development, should be made a fundamental commitment for all people of goodwill and common sense, all who want to protect our western culture and ethics, all who care about the future of humanity.[10]

Rita Levi-Montalcini, MD (1909–2012)

Rita Levi-Montalcini currently holds the record as the longest-lived Nobel Laureate. When she died in Rome, Italy, on 30 December 2012, she had lived for 103 years and 252 days. Her longevity meant she outlived many of the most important figures in her life, including her mother, Adele Levi née Montalcini, in 1963, her brother Gino in 1974, and both her twin sister Paola and elder sister Anna in 2000. Her mentor, Giuseppe Levi, died in 1965 at the age of 92. Rita visited him in Turin two weeks before his death, where they 'reminisced about what we had enjoyed and suffered together in the past' and said goodbye.[11] Her longtime colleague and friend at Washington University, Viktor Hamburger, died age 100 in 2001. And her co-Nobel Laureate Stanley Cohen died in 2020 at age 97.

At least officially, Rita retired in 1979 but remained professionally and personally active long after. In some ways, her life became even more active after her retirement. It was also the period during which she finally achieved full recognition for her scientific accomplishments. When Rita won the Nobel Prize in 1986, she was 77 years old. She filled her remaining quarter century with her scientific work and many other pursuits besides. She wouldn't have had it any other way. As she said in an interview published in honour of her 100th birthday in 2009, 'It's not enough what I did in the past – there is also the future.'[12] She wrote prolifically, including about her life and other topics close to her heart, such as ageing, the brain, and African women's education.[13] Her notoriety also helped her publish previously written works, like one from 1971, titled 'The Feminine Awakening' about the history of women's emancipation up to 1970. Perhaps because she was so open about her Holocaust experiences and they were so well known, she did not give a dedicated Holocaust testimony, at least that I am aware of. However, she did talk publicly about her experiences. According to one account, written during Rita's life:

> Rita speaks to adolescents in schools throughout Italy, urging them to face life with optimism and faith in people, and quoting Ann [sic] Frank's farewell works on the eve of the discovery of the Frank hiding place by the Nazis: 'I think that this cruelty too will end, that peace and tranquility will return again. In the meantime I must uphold my ideals, and perhaps the time will come when I shall be able to carry them out.' It came for Rita.[14]

As a dual American and Italian citizen, Rita was active in both countries. In addition to founding the Centre of Neurobiology in Rome in the early 1960s, she also founded the European Brain Research Institute in Rome in 2006 as a multidisciplinary research centre for young scientists, as well as the charitable organisation the Rita Levi-Montalcini Foundation in 2001. In August 2001, she was appointed as Senator for Life in Italy by the Italian President Carlo Azeglio Ciampi. She did not take this merely as an honorary position but was politically active and vocal. She especially supported women in science, both generally and on an individual level.[15]

In the introduction to her autobiography, *In Praise of Imperfection*, Rita explained the title and summarised her personal philosophy in life, which makes a nice epitaph:

> The fact that the activities that I have carried out in such imperfect ways have been and still are for me a source of inexhaustible joy, leads me to believe that imperfection, rather than perfection, in the execution of our assigned or elected tasks is more in keeping with human nature.[16]

Halina Maria Czajkowska Robinson (1927–2021)

Halina Czajkowska Robinson was 95 when she died on 27 September 2021, in London, Ontario, Canada, during the COVID-19 pandemic. Her husband, Russell 'Rus' Robinson, predeceased her, dying on 10 April 2010, at age 90. They had been married for 57 years and had no children, which was a great sadness to Halina.[17] Halina's mother, Emilia Czajkowska (née Michalowska), died age 82 in 1981, after a long struggle with poor health and PTSD. Halina's father, Czesław Karol Czajkowski, lived with Halina and Rus after Emilia's death, and died at the advanced age of 104.[18] Czesław, Emilia, Halina, and Rus are all buried in St Peter's Cemetery in London, Ontario, where a Polish Military Monument stands in memory of Polish soldiers who fought in the Second World War.[19] Czesław had been an officer in the Polish military and fought for Poland as a colonel during the Second World War. Rus, though not Polish, had also served in the Second World War.[20]

Aside from her memoir, *Heaven Hell, and Purgatory: A Canadian Memoir of a Happy Polish Childhood, Nazi Horror, and Swedish Refuge* (2020), Halina did not appear to have given any recorded interviews about her experiences during the Second World War. Explaining her motivation for writing her memoir, she wrote, 'The good part is for my pleasure, the bad part is in remembrance of those who had the same hard, painful experiences, but who could not tell them for one reason or another.'[21] I was fortunate to play a small part in Halina's memoir after I was contacted in 2020 by Jacalyn Duffin regarding Halina's experiences in Sweden.

When Halina retired in 1977, she 'resumed her longstanding interest in the fine arts', according to Duffin.[22] After her death, it was discovered that she left behind a trove of around 1,000 works of her own art, framed and curated. In 2023, some of this art was exhibited in London, Ontario.[23] In addition to this legacy, Halina supported the library collection established by Rus in 1979 at Western University after his death. After her own death, an endowment of two million Canadian dollars (CAD) – its largest donation to that date – in Halina and Rus's names was made to Western Libraries for books and resources in military and veterans' history.[24]

Eva Klein, MD, PhD (1925–2025)

Eva Klein was the last of the six women to pass away, dying in Stockholm, Sweden, just four days shy of her 100th birthday on 19 January 2025. Georg predeceased her, passing away at the age of 91 on 10 December 2016. In 2015, Eva and Georg were featured in the Swedish national newspaper *Dagens Nyheter*, not for their scientific accomplishments but for each reaching the age of 90. They had been married for 69 years at that point and had been living in the same house just outside of Stockholm since 1960. Both went into their offices at KI four times a week.[25] 'Work keeps me young,' Eva told a Swedish radio journalist later that year.

> Eva Klein has been researching cancer her whole life. Now she is 90 and has no plans to stop. She has her workplace in an office at Karolinska Institutet. Her husband sits in the next room. 'But we try not to talk too much about our work. He doesn't want that,' says Eva, who believes that work is the key to her still being active. 'You have to have something that is the meaning of life, to be able to do something that is important and that has meaning.'[26]

She continued her work at KI until Georg's death. 'For those who had the privilege of being her students and colleagues, Eva was very special: creative, insightful and warm, with a deep understanding of human nature, outspoken, unconventional to almost defiant, and with a profound sense of humor,' two colleagues wrote of Eva after her death. 'Science was, for her, "above all great fun."'[27] To honour Eva and Georg on their 80th birthdays in 2005, their former students established the Georg and Eva Klein Foundation 'to support research in the fields of immunology, cell biology, infections and cancer in "the spirit of Georg and Eva"'.[28] After an absence of three decades, Eva began visiting Hungary in the 1970s in a professional capacity and gradually developed close professional ties with Hungarian scientists. She subsequently received many honours and awards in Hungary. 'The Klein spirit is characterized by an intensive international exchange of knowledge, a major commitment to young researchers, a lack of organizational hierarchy and a perspective that encompasses both the natural sciences and the humanities,' proclaimed the Hungarian Academy of Sciences in 2014.[29] Eva appreciated the respect she was given in Hungary, particularly by non-Jews. Considering what had happened there in her youth, it is understandable that this was so important to her.[30]

Eva's father, Josef Fischer, died in Hungary in 1948, the same year she left for Sweden. Her mother, Iren Fischer, also died in Hungary in 1962. Eva saw her just before she died after many years apart. Laszlo Fischer, Eva's brother, also settled in Stockholm, Sweden, where he died in 1993. Her sister, Veronika Lengyel, remained in Hungary. János (Jan) Szirmai (1925–2014), the non-Jewish Hungarian friend who hid Eva and other Jews in Budapest, and Dr Clara Ambrus, née Bayer (1924–2011), Eva's non-Jewish friend who let Eva use her identity papers, were recognised by Yad Vashem as Righteous Among the Nations – a non-Jewish person who risked their life to save Jews from the Nazis – in 2006.[31]

Although Eva was alive while I was researching for and writing this book, I never made any attempt to contact her due to her age and my feeling that she deserved to be left in peace. At times, I wavered in my resolve and when I learned she passed away while I was writing this book, I recriminated with myself for not reaching out to her. It feels appropriate to end this book with her, the last of the six STEM pioneers to pass away.

Notes

Introduction

1. My earliest published writing on the subject was: Victoria Martínez, 'History dies deep in the woods: The forgotten Nazi concentration camp survivors in the forests of Småland', *The Local Sweden*, 4 July 2017, www.thelocal.se/20170704/history-dies-deep-in-the-woods-the-forgotten-nazi-concentration-camp-survivors-in-the-forests-of-smaland-sweden. Accessed 4 October 2024.
2. 'Remaining Jewish Population of Europe in 1945', *Holocaust Encyclopedia*, United States Holocaust Memorial Museum, https://encyclopedia.ushmm.org/content/en/article/remaining-jewish-population-of-europe-in-1945. Accessed 8 August 2024.
3. Ruth Feldman, 'Women in Science: myth, harsh reality, or advantage', *Frontiers in Human Neuroscience* 17 (2023). Matthew B. Ross et al., 'Women are credited less in science than men', *Nature* 608 (4 August 2022). 'Women totalled almost a third of STEM graduates in 2021', eurostat, 8 March 2024, https://ec.europa.eu/eurostat/web/products-eurostat-news/w/ddn-20240308-2. Accessed 9 August 2024.
4. e.g. Lisa Yount, *Rita Levi-Montalcini: Discoverer of Nerve Growth Factor (Makers of Modern Science).* Chelsea House Publishing, 2009. Francesca Valente, *Rita Levi-Montalcini: Pioneer & Ambassador of Science*. Barbera Foundation, 2021.
5. Halina Maria Czajkowska Robinson, *Heaven. Hell, and Purgatory: A Canadian Memoir of a Happy Polish Childhood, Nazi Horror, and Swedish Refuge.* Jacalyn Duffin, ed. Kingston, Ontario, Canada: Poison Spindle Press, 2020. Jacalyn Duffin, 'Poisoning the Spindle: Serendipity and Discovery of the Anti-Tumor Properties of Vinca Alkaloids (Part I)', *Pharmacy in History* 44, 2 (2002): 64–76. Jacalyn Duffin, 'Poisoning the Spindle: Serendipity and Discovery of the Anti-Tumor Properties of Vinca Alkaloids (Part II)', *Pharmacy in History* 44, 3 (2002): 105–118. Jacalyn Duffin, 'Epilogue', *Pharmacy in History* 44, 3 (2002): 118–119.
6. Roger A. Ritvo & Diane M. Plotkin, *Sisters in Sorrow: Voices of Care in the Holocaust.* College Station: Texas A&M Press, 1998.

Prologue

1. Eva Klein, Interview 32450. Interview by Jan Walter Levy. Visual History Archive, USC Shoah Foundation, 30 August 1997. https://vha.usc.edu/testimony/32450; Georg Klein, Interview 32435. Interview by Jan Walter Levy. Visual History Archive, USC Shoah Foundation, 30 August 1997. https://vha.usc.edu/testimony/32435.
2. Osbert Sitwell, Introduction to *Left Hand, Right Hand! An Autobiography, Vol. 1: The Cruel Month.* London: Macmillan & Co, Ltd, 1945, p. vi.
3. 'Our History', Leukemia & Lymphoma Society, www.lls.org/our-history. Accessed 7 August 2024.

4. 'History, From 2 AD to the present', National Bleeding Disorders Foundation, www.bleeding.org/bleeding-disorders-a-z/overview/history. Accessed 7 August 2024.
5. Jess Jawin, 'Solidarity Day', *Texas Jewish Post*, vol. XXXIV, no. 18, 1 May 1980, p. 1.
6. Norwood Oakley Hill. *Blood and Betrayal: The Untold Story*. Self-published, 2019, p. 63.
7. 'The Memoirs of Ellen Loeb', United States Holocaust Memorial Museum (USHMM), RG-50.751*0016, pp. 21–22.
8. Hill, *Blood and Betrayal*, op cit., p. 62.
9. 'The Memoirs of Ellen Loeb', op cit., p. 18.
10. Hill, *Blood and Betrayal*, p. 62.
11. 'Hemophilia', World Federation of Hemophilia, https://elearning.wfh.org/elearning-centres/hemophilia. Accessed 9 August 2024. 'Blood cancer – everything you need to know', Worldwide Cancer Research, www.worldwidecancerresearch.org/news-opinion/2022/september/blood-cancer-everything-you-need-to-know. Accessed 7 August 2024. 'Childhood cancer: Key facts', World Health Organization, 13 December 2021, www.who.int/news-room/fact-sheets/detail/cancer-in-children. Accessed 7 August 2024.
12. See, for example: 'Top 10 causes of death in the U.S. in 1900 and 2022', Statista.com, www.statista.com/statistics/235703/major-causes-of-death-in-the-us. 'Achievements in Public Health, 1900–1999: Control of Infectious Diseases', CDC.gov, www.cdc.gov/mmwr/preview/mmwrhtml/mm4829a1.htm. 'Causes of death over 100 years', ONS.gov.uk, www.ons.gov.uk/peoplepopulationandcommunity/birthsdeathsandmarriages/deaths/articles/causesofdeathover100years/2017-09-18. Accessed 14 June 2024.
13. Freddie Bray et al., 'Global cancer statistics 2022: GLOBOCAN estimates of incidence and mortality worldwide for 36 cancers in 185 countries', *CA: A Cancer Journal for Clinicians* 74, 3 (2024): 229–263, p. 229. 'Global cancer burden growing, amidst mounting need for services', World Health Organization, 1 February 2024, www.who.int/news/item/01-02-2024-global-cancer-burden-growing--amidst-mounting-need-for-services. Accessed 14 June 2024.
14. Stanley Plotkin, 'History of vaccination', *PNAS* 111, 34 (2014): pp. 12283–12287. Rino Rappuoli, 'Vaccines: Science, health, longevity, and wealth', *PNAS* 111, 34 (2014): 12282. Seamus McAleer, 'A history of cancer and its treatment', *Ulster Medical Journal* 91, 3 (2022): 124–129, pp. 124–125.
15. Gulten Dinc and Yesim Isil Ulman, 'The introduction of variolation "A La Turca" to the West by Lady Mary Montagu and Turkey's contribution to this', *Vaccine* 25, 21 (2007): 4261–4265.
16. Jérôme Galon and Daniela Bruni, 'Tumor Immunology and Tumor Evolution: Intertwined Histories', *Immunity* 52 (2020): 55–81. Christopher R. Parish, 'Cancer immunotherapy: The past, the present and the future', *Immunology and Cell Biology* 81 (2003): 106–113.
17. George Klein and Eva Klein, 'How One Thing Has Led to Another', *Annual Review of Immunology* 7 (1989): 1–33, p. 16.
18. Ingemar Ernberg, 'George Klein (1925–2016)', *Nature* 542, 16 February 2017: 296.
19. R. Kiessling, E. Klein, H. Pross, and H. Wigzell, '"Natural" killer cells in the mouse. II. Cytotoxic cells with specificity for mouse Moloney leukemia cells. Characteristics of the killer cell', *European Journal of Immunology* 5, 2 (1975): 117–121. 'Spotlight on: NK Cells', Karolinska Institute, https://ki.se/en/research/popular-science-and-dialogue/spotlight-on/spotlight-on-nk-cells. Accessed 16 June 2024.

20. Alireza Zafarani et al., 'Natural killer cells in COVID-19: from infection, to vaccination and therapy', *Future Virology* 2023.
21. Parish, 'Cancer immunotherapy', op cit., p. 108.
22. See, for example: Gavin P. Dunn, 'Cancer Vaccines: Preventive, Therapeutic, Personalized', Cancer Research Institute, www.cancerresearch.org/treatment-types/cancer-vaccines. Accessed 19 June 2024.
23. See, for example: Andrew Gregory, 'What are cancer vaccines and have scientists finally found a cure?' *The Guardian*, www.theguardian.com/science/article/2024/may/31/what-are-cancer-vaccines-and-have-scientists-finally-found-a-cure. Accessed 19 June 2024.
24. Susan L. Smith, 'War! What is it good for? Mustard gas medicine', *Canadian Medical Association Journal* 189, 8 (2017): E321–E322.
25. 'From the field of battle, an early strike at cancer', *Yale Medicine Magazine*, 2005, https://medicine.yale.edu/news/yale-medicine-magazine/article/from-the-field-of-battle-an-early-strike. Accessed 29 May 2024.
26. Duffin, 'Poisoning the Spindle (Part II)', op cit., p. 112.
27. Robert L. Noble, 'The Discovery of the Vinca Alkaloids-Chemotherapeutic Agents against Cancer', *Biochemistry and Cell Biology* 68 (1990): 1344–1351.
28. Duffin, 'Poisoning the Spindle (Part I)', op cit., pp. 68–69. Duffin, Part II, op cit., p. 110.
29. 'Vinblastine and vincristine: life-saving drugs from a periwinkle', John Innes Centre, www.jic.ac.uk/blog/vinblastine-and-vincristine-life-saving-drugs-from-a-periwinkle. Accessed 29 May 2024. Aliyah Kovner, 'An Anti-cancer Drug in Short Supply Can Now be Made by Microbes', Berkeley Lab, 31 August 2022. https://newscenter.lbl.gov/2022/08/31/anti-cancer-drug-now-made-by-microbes. Accessed 29 May 2024.
30. Berkeley Lovelace, 'Pediatric cancer drugs in shortage as drug supply crisis drags on', *NBC News*, 25 September 2023. www.nbcnews.com/health/health-news/childrens-cancer-drugs-shortage-supply-crisis-continues-rcna105854. Accessed 29 May 2024.
31. Kovner, 'Anticancer drug in short supply', op cit. Justin E. Sears and Dale L. Boger. 'Total Synthesis of Vinblastine, Related Natural Products, and Key Analogues and Development of Inspired Methodology Suitable for the Systematic Study of Their Structure–Function Properties', *Accounts of Chemical Research* 48, 3 (2015): 653–662.
32. See, e.g. Paul J. Carrick, *Medical Ethics in the Ancient World*. Washington, DC: Georgetown University Press, 2001, p. 15. Burton R. Anderson, 'Comparison of ancient Mesopotamian and Hippocratic medicine', *Hektoen International, A Journal of Medical Humanities*, 2, 1 (2010). https://hekint.org/2017/01/22/comparison-of-ancient-mesopotamian-and-hippocratic-medicine. Accessed 13 August 2024.
33. See, e.g. Teja Lele, 'India's ancient tradition that aligns mind, body and spirit – and where to experience it', BBC Travel, 18 April 2024. www.bbc.com/travel/article/20240417-ayurveda-indias-ancient-tradition-that-aligns-mind-body-and-spirit. Accessed 13 August 2024.
34. See, e.g. Faith Lagay, 'The Legacy of Humoral Medicine', *Virtual Mentor* 4, 7 (2002): 206–208.
35. See, e.g. Ian M. Handley et al., 'Quality of evidence revealing subtle gender biases in science is in the eye of the beholder', PNAS 112, 43 (2015): 13201–13206, pp. 13201–13206. Rita Colwell and Sharon Bertsch McGrayne, *A Lab of One's Own: One Woman's*

Journey Through Sexism in Science. New York: Simon & Schuster Paperbacks, 2020, pp. 181–193.

36. Anthony Liversidge, 'Interview: Rita Levi-Montalcini', *Omni*, March 1988, p. 72.
37. Liversidge, 'Interview: Rita Levi-Montalcini', p. 72.
38. Rita Levi-Montalcini, Trans. Luigi Attardi, *In Praise of Imperfection: My Life and Work.* New York: Basic Books, 1988, pp. 141–142.
39. See, e.g. Luigi Aloe et al., 'Nerve growth factor: from early discoveries to the potential clinical use', *Journal of Translational Medicine* 10, 239 (2012): 1–15.
40. Attributed to neuroscientist A. D. Smith, 1986, and cited in Luigi Aloe, 'Rita Levi-Montalcini: A Brief Biographic View of Past and Present Studies on Nerve Growth Factor', *Microscopy Research and Technique* 45 (1999): 208–209, p. 208.
41. On B12, see, e.g. John M. Scott, Anne M. Molloy, 'The Discovery of Vitamin B12', *Annals of Nutrition & Metabolism* 61, 3 (2012): 239–245. 'The Vitamin B Complex: A National Historic Chemical Landmark', ACS, www.acs.org/education/whatischemistry/landmarks/vitamin-b-complex.html. Accessed 12 September 2024.
42. SOU (Swedish Government Official Reports) 1995: 110. *Viljan att veta och viljan att förstå : kön, makt och den kvinnovetenskapliga utmaningen i högre utbildning : slutbetänkande* [trans. The will to know and the will to understand: gender, power and the feminist challenge in higher education: final report], p. 226. https://lagen.nu/sou/1995:110. Accessed 13 September 2024. On Dorothy Hodgkin, see Georgina Ferry, *Dorothy Hodgkin: A Life.* Pennsylvania: Cold Spring Harbor Laboratory Press, 2000.
43. e.g. Halina Y. Neujahr and H. Lundin, 'On Vitamins in Sewage Sludge. I. Vitamin B12, Folic Acid, and Folinic Acid Factors in Digested Municipal Sewage Sludge', *Acta Chemica Scandinavica* 7, (1953): 1036–1040. Following this first publication, she and Lundin published nine further articles as a continuation.
44. Halina Neujahr, Interview 30476. Interview by Jan Walter Levy. Visual History Archive, USC Shoah Foundation, 15 April 1997. https://vha.usc.edu/testimony/30476, tapes 7 & 8. SOU 1995:110, p. 225.
45. See, e.g. E. Lester Smith, 'Isolation and Chemistry of Vitamin B12', Chapter 1 in *The Biochemistry of Vitamin B12, Biochemical Society Symposia No. 13*, edited by R. T. Williams. Cambridge University Press, 1955, p. 4.
46. 'Cardiovascular diseases', World Health Organization, www.who.int/health-topics/cardiovascular-diseases. Accessed 20 May 2024.
47. Ruth Pick, 'Ruth Pick to Horace Davenport', *The Physiologist*, February 1979, p. 32.
48. Lillian Pompian, 'Your Heart is in Their Work', *Chicago Sunday Tribune* (Chicago, Illinois), 24 May 1959, p. 22.
49. 'Woman Doctor Plans Speech', *The De Kalb Daily Chronicle* (De Kalb, Illinois), 12 June 1962, p. 2. 'Ruth Pick To Speak at Meeting of Heart Units', *Sycamore Tribune* (Sycamore, Illinois), 8 June 1962, p. 7.
50. 'Aspirin May Halt Strokes', UPI, *The Atlanta Constitution* (Atlanta, Georgia), 3 May 1979, p. 9-A. Ruth Pick, Interview 26278. Interview by Sandy Rebitzer. *Visual History Archive*, USC Shoah Foundation, 21 February 1997, tape 5. https://vha.usc.edu/testimony/26278
51. Jeffrey S. Berger, 'Aspirin for Primary Prevention – Time to Rethink Our Approach', *JAMA Network Open* 5, 4 (2022), https://jamanetwork.com/journals/jamanetworkopen/fullarticle/2791538. Accessed 21 May 2024.
52. Pick, USC interview, tape 5.

53. Laila Aryan et al., 'The Role of Estrogen Receptors in Cardiovascular Disease', *International Journal of Molecular Sciences* 21, 12 (2020): 4314.
54. For further reading on this topic, see, for example, Colwell, *A Lab of One's Own*, op cit., especially pp. 181–193.
55. Allen B. Weisse, *Heart to Heart: The Twentieth-Century Battle Against Cardiac Disease*. New Brunswick, New Jersey, and London: Rutgers University Press, 2002.
56. Ibid., p. 281. Emphasis added.
57. Matthew B. Ross et al., 'Women are credited less in science than men', *Nature* 608, 4 August 2022.
58. Ross D. Feldman, 'Sex-Specific Determinants of Coronary Artery Disease and Atherosclerotic Risk Factors: Estrogen and Beyond', *Canadian Journal of Cardiology* 36 (2020): 706–711, p. 706.
59. Lea Merone et al., 'Sex Inequalities in Medical Research: A Systematic Review of the Literature', *Women's Health Reports* 3, 1 (2022): 29–59.

Chapter 1

1. Pick, 'Ruth Pick to Horace Davenport', op cit.
2. Suggested reading: Margaret W. Rossiter, *Women Scientists in America, Volume 2, Before Affirmative Action, 1940–1972*, Baltimore & London: The Johns Hopkins University Press, 1995. Amelia Bonea and Irina Nastasă-Matei, eds., *Negotiating in/Visibility: Women, Science, Engineering and Medicine in the Twentieth Century.* Manchester: Manchester University Press, 2025.
3. Levi-Montalcini, *In Praise of Imperfection*, op cit., pp. 149–150.
4. Rita Levi-Montalcini, 'Effects of Mouse Tumor Transplantation on the Nervous System', *Annals of the New York Academy of Sciences* 55 (1952): 330–344.
5. Giovanni Giudice, 'From a home-made laboratory to the Nobel Prize: An interview with Rita Levi-Montalcini', *International Journal of Developmental Biology* 44 (2000): 563–566, p. 564. To be fair, she also writes in her autobiography that she 'was congratulated by the most important personalities present', so it wasn't a complete disappointment. See Levi-Montalcini, *In Praise of Imperfection*, op cit., pp. 151–152.
6. To the best of my knowledge, no book-length biographies of Hertha Meyer exist.
7. Levi-Montalcini, *In Praise of Imperfection*, p. 154.
8. Ibid., p. 163.
9. Alison Abbott, 'One Hundred Years of Rita', *Nature* 458, 7328 (May 2009): 564–567, p. 566. The biochemist quoted is Ralph Bradshaw.
10. Rossiter, *Women Scientists*, op cit., pp. 50–68.
11. See, e.g. Rossiter, *Women Scientists*, pp. 63–67.
12. Rossiter, *Women Scientists*, pp. 98–110.
13. John Barbour, 'Ingredient in Mice Saliva Aids Control Nerve Growth', *The Palladium-Item* (Richmond, Indiana), 19 November 1959, p. 21. Thomas R. Henry, 'Vistas in Science: Snake Venom Promotes Nerve Growth', *Evening Star* (Washington, DC), 20 May 1960, p. 22.
14. 'A Celebration of Women at Washington University in St. Louis', *Timeline of Faculty Leaders, 1953–2010*, University Archives, Washington University Archives, https://uaexhibits.omeka.net/exhibits/show/celebrating women-washington/faculty/timeline-of-faculty-leaders--2. Accessed 17 August 2024. The site implies that the year of her promotion to full professor was 1953. In fact, this took place in 1958. Regarding female

professors of Zoology, see Rossiter, *Women Scientists*, p. 129, which includes a table showing that in 1960, none of the twenty leading universities in the US had a woman who was a full professor in Zoology.

15. George Barker, 'Super Dooper Snooper', *The Nashville Tennessean Magazine* (Nashville, Tennessee), 18 March 1962, pp. 12–13, 22.
16. See, e.g. 'Dr. Levi-Montalcini To Receive Award', *St. Louis Post-Dispatch* (St. Louis, Missouri), 8 March 1963, p. 3B. 'Dr. Levi-Montalcini Receives Award', *St. Louis Post-Dispatch*, 29 November 1963, p. 3B.
17. Abbot, 'One Hundred Years of Rita', op cit., p. 566.
18. Levi-Montalcini, *In Praise of Imperfection*, p. 168.
19. This is the title of a book she wrote: Rita Levi-Montalcini, *The Saga of the Nerve Growth Factor: Preliminary Studies, Discovery, Further Development*, World Scientific Series in 20th Century Biology. Istituto di Neurobiologia, Italy, 1997.
20. Liversidge, 'Interview: Rita Levi-Montalcini', op cit., p. 72.
21. Levi-Montalcini, *In Praise of Imperfection*, p. 94.
22. Ibid., p. 91.
23. Ibid., p. 97.
24. See, e.g. 'Does a Nursing Career offer me the things I want in life?' *Green Bay Press-Gazette* (Wisconsin, Illinois). 28 December 1949, p. 16.
25. 'The Saturday Evening Post', *St. James Encyclopedia of Popular Culture.* Encyclopedia. com. www.encyclopedia.com/media/encyclopedias-almanacs-transcripts-and-maps/saturday-evening-post. Accessed 15 May 2024.
26. Steven M. Spencer, 'Do Women Make Good Doctors?' *The Saturday Evening Post*, Philadelphia, Penn., 221, 29, 13 November 1948.
27. William Weinfeld, 'Income of Physicians, 1929–1949', *Survey of Current Business*, July 1951. Ellen S. More and Marilyn J. Greer, 'American Women Physicians in 2000: A History in Progress', *JAMWA* 55, 1 (2000): 6–9.
28. Wienfeld, 'Income of Physicians', p. 24.
29. Pick, USC interview, op cit., tape 4.
30. Spencer, 'Do Women Make Good Doctors?' op cit., p. 16.
31. Pick, 'Ruth Pick to Horace Davenport', op cit.
32. Sarah Gordon, ed. *All Our Lives: A Centennial History of Michael Reese Hospital and Medical Center, 1881-1981.* Chicago: Michael Reese Hospital and Medical Center, 1981, pp. 119-120.
33. 'Our History', Michael Reese, https://wearemichaelreese.org/history. Accessed May 18, 2024.
34. Cutler, *The Jews of Chicago*, op cit., pp. 356–358.
35. Pick, USC testimony, tape 4.
36. Ibid., tape 5.
37. Pompian, 'Your Heart', op cit., p. 22.
38. Ibid., p. 24.
39. Weisse, *Heart to Heart*, p. 283. Pick, USC testimony, tape 5.
40. Quoted in Weisse, *Heart to Heart*, p. 283.
41. 'Lillian Pompian, 83: Freelanced for Tribune', *Chicago Tribune*, 18 August 2000.
42. Pompian, 'Your Heart', 22.
43. 'Death rate from cardiovascular diseases, 1950 to 2021', Our World in Data, https://ourworldindata.org/grapher/cardiovascular-disease-death-rate-who-mdb. Accessed 29 May 2024.

44. James E. Dalen, 'The Epidemic of the 20th Century: Coronary Heart Disease', *The American Journal of Medicine* 127, 9 (2014): 807–811.
45. Pick, USC testimony, tape 5.
46. H.J. 'Svenskar löste cancerproblem', *Svenska Dagbladet*, 17 March 1954, pp. 1 and 7. Translated from Swedish.
47. Eva Klein, Interview 32450. Interview by Jan Walter Levy. *Visual History Archive*, USC Shoah Foundation, 30 August 1997, tape 5. https://vha.usc.edu/testimony/32450.
48. H. J., 'Svenskar löste cancerproblem', p. 7.
49. Eva Klein, *Transformation of solid into ascites tumors.* Dissertation, Stockholm: Karolinska Institutet, 1955.
50. Klein, USC Interview, op cit., tape 5.
51. George Klein and Eva Klein, 'Tumor Immunology', *Immunology: The Making of Modern Science*, Richard B. Gallagher, et al. eds. Cambridge, Mass: Academic Press Ltd, 1995, p. 204.
52. Jérôme Galon and Daniela Bruni, 'Tumor Immunology and Tumor Evolution: Intertwined Histories', *Immunity* 52 (2020): 55–81. Parish, 'Cancer immunotherapy', op cit., pp. 106–113.
53. Klein and Klein, 'How One Thing Has Led to Another', op cit., p. 16.
54. Elianne Riska, 'Towards gender balance: but will women physicians have an impact on medicine?' *Social Science &Medicine* 52, 2 (2001): 179–187, p. 181.
55. Alina M. Lindegren, *Education in Sweden*. Bulletin 1952, No. 17. Washington, DC: United States Government Printing Office, 1952, p. 44.
56. Klein and Klein, 'How One Thing Has Led to Another', p. 13.
57. Patrick McGee, 'Shattering the Glass Beaker: Women's Past and Present Have Played Key Roles at Fox Chase', *Fox Chase Forward*, Winter/Spring 2020, pp. 6–13.
58. Klein, USC interview, tape 5.
59. Klein and Klein, 'How One Thing Has Led to Another', pp. 11–15.
60. Klein and Klein, 'How One Thing Has Led to Another', pp. 11–15.
61. She credits the mice in a journal article published in 1955: Eva Klein, 'Immediate Transformation of Solid into Ascites Tumors. Studies on a Mammary Carcinoma of an Inbred Mouse Strain', *Experimental Cell Research* 8 (1955), 213–225, p. 213.
62. H. J., 'Svenskar löste cancerproblem', p. 7.
63. Klaus Rajewsky, 'George Klein: 1925–2016', *PNAS* 114, 13 (2017): pp. 3275–3277.
64. Everett Taylor Jr., 'Huntin' 'n Peckin'', *The Tyler Courier-Times* (Tyler, Texas), 17 July 1960, p. 1.
65. Katricia Cochran, 'Four-Year-Old Named Charity Game Mascot', *Wichita Falls Times* (Wichita Falls, Texas), 1 July 1960, p. 8.
66. Katricia Cochran, 'Final Rites are Held for Leukemia Victim', *Wichita Falls Times*, 6 September 1960, p. 2.
67. Preston McGraw, 'MD Escapes Gas Chambers, Devotes Life to Research', *Leader-Telegram* (Eau Claire, Wisconsin), 25 March 1961, p. 4.
68. Society [section]. 'To Speak Here Friday', *McKinney Daily Courier-Gazette* (McKinney, Texas), 3 March 1964, p. 4.
69. Loeb, 'The Memoirs of Ellen Loeb', op cit., p. 17.
70. Ibid., p. 19.
71. Ibid., pp. 21–22.

72. Marc Perry et al., 'National Archives Will Soon Release 1950 Census, Offering a Look at a Defining Time in Modern U.S. History', United States Census Bureau, www.census.gov/library/stories/2022/03/1950-census-records-window-to-history.html. Accessed 18 June 2024.
73. W. W. Bauer, 'Varied Fields of Medicine Drawing Greater Number of Women Doctors', *Fort Worth Star-Telegram* (Fort Worth, Texas), 11 December 1953, p. 11.
74. Eleanor Kerstetter, 'Career and Motherhood Combined', *Corpus Christi Times* (Corpus Christi, Texas), 12 April 1959, p. 41.
75. George W. Crane, 'The Worry Clinic: She Prefers Male Bosses', *Wichita Falls Times* (Wichita Falls, Texas), 27 July 1962, p. 20.
76. Bill Thompson, 'The Brave World of Leukemia Victims, And Hope Offered by Wadley Center…', *The Paris News* (Paris, Texas), 25 February 1960, p. 4.
77. See, for example: Britt P. Tevis, '"Jews Not Admitted": Anti-Semitism, Civil Rights, and Public Accommodation Laws', *The Journal of American History*, March 2021, pp. 847–870. Bernie Farber, 'No Dogs and Jews Allowed', *HuffPost Canada*, 13 May 2013, www.huffpost.com/archive/ca/entry/no-dogs-and-jews-allowed_b_1510399. Accessed 19 June 2024.
78. On the history of Jews in Texas see, e.g. Hollace Ava Weiner & Kenneth D. Roseman, eds. *Lone Stars of David: The Jews of Texas*, University Press of New England, Lebanon, New Hampshire: 2007.
79. Laurel Leff, 'Combating Prejudice and Protectionism in American Medicine: The Physicians Committee's Fight for Refugees from Nazism, 1939–1945', *Holocaust and Genocide Studies* 28, 2 (2014): pp. 181–239.
80. Lanie Francis, 'Women in Hematology – Where Are We Today?' *The Hematologist* 4, 3 (2007).
81. Society. 'To Speak Here Friday', op cit.
82. See, e.g. Joseph M. Hill and Ellen Loeb, 'Treatment of Leukemia, Lymphoma and Other Malignant Neoplasms with Vinblastine', *Cancer Chemotherapy Reports* 15 (1961): 41–61. S. K. Carter, 'Cisplastin – Past, Present, and Future', *Platinum Coordination Complexes in Cancer Chemotherapy*, Boston: Martinus Nijhoff Publishing, 1984, p. 359.
83. See, e.g. J. M. Hill and Ellen Loeb, 'Autologous Bone Marrow Replantation', *Experimental Hematology* 5 (1962): 23.
84. Hearing Before the Subcommittee on Immigration and Claims of the Committee on the Judiciary, House of Representatives, One Hundred Fourth Congress, Second Session on H.R. 1023, Ricky Ray Hemophilia Relief Fund Act of 1995, September 19, 1996, Serial No. 122, U.S. Government Printing Office, Washington: 1997, 259–260.
85. 'Kvinna för sin hatt', *Stockholms-Tidningen* (Stockholm, Sweden), 31 May 1960, p. 1. The same photo and caption also appeared in, for example, the southern Swedish newspaper, *Sölvesborgstidningen* on 2 June 1960, p. 6.
86. Halina Neujahr, Interview 30476. Interview by Jan Walter Levy. Visual History Archive, USC Shoah Foundation, 15 April 1997. https://vha.usc.edu/testimony/30476, tape 7. Accessed 13 September 2024.; SOU [Swedish Official Government Reports] 1995:10, *Viljan att veta och viljan att förstå : kön, makt och den kvinnovetenskapliga utmaningen i högre utbildning : slutbetänkande*, pp. 222–223; Katarina Rydberg, 'Forskare mot all odds', *KTH-nytt* 4/93, p. 9.
87. Neujahr, USC interview, tape 2. SOU 1995: 110, op cit., p. 220.
88. SOU 1995:110, pp. 223–224.

89. Ibid., p. 224.
90. Neujahr, USC interview, tape 5. Jerzy Adamczyk and Carlösten Nordmark, 'Hon började sin forskarbana i gettot', *Aftonbladet* (Stockholm, Sweden), 25 April 1971, p. 17.
91. SOU 1995:110, p. 224.
92. Halina Neujahr, 'Bioteknik', 2 May 1983, Halina Neujahrs Arkiv, Nordiska Museet, Stockholm, 6754:1/F 1 a/1.
93. SOU 1995:110, p. 224.
94. Ibid.
95. Ibid., pp. 226–228.
96. Rydberg, 'Forskare mot all odds', op cit., p. 9. Translated from Swedish.
97. Ibid. See also the Professional CV of Halina Neujahr, dated 14 October 2000, Nordiska Museet, Stockholm, 6754:1/F 1 a/1. Among Halina's published articles on this line of research are: J. M. Varga and H. Y. Neujahr, 'Isolation from soil of phenol-utilizing organisms and metabolic studies on the pathways of phenol degradation', *Plant Soil* 33: 565–571 (1970). H. Y. Neujahr and A. Gaal (1973), 'Phenol Hydroxylase from Yeast', *European Journal of Biochemistry*, 35: 386–400.
98. See, e.g. Halina Y. Neujahr, 'Biosensors for environmental control', *Biotechnology and Genetic Engineering Reviews* 1.1 (1984): 167–186.
99. Duffin, 'Poisoning the Spindle (Part II)', op cit., p. 110.
100. Katy Kelleher, 'Periwinkle, the Color of Poison, Modernism, and Dusk', *The Paris Review*, 19 August 2020. www.theparisreview.org/blog/2020/08/19/periwinkle-the-color-of-poison-modernism-and-dusk. Accessed 18 May 2024.
101. Nagmeh Nejat et al., 'Ornamental Exterior versus Therapeutic Interior of Madagascar Periwinkle (*Catharanthus roseus*): The Two Faces of a Versatile Herb', *The Scientific World Journal*, 2015, p. 1.
102. Duffin, 'Poisoning the Spindle (Part I)', op cit., pp. 68–69; Duffin, 'Part II', op cit., p. 110.
103. Duffin, 'Poisoning the Spindle (Part II)', p. 110. See also Jacalyn Duffin, 'Editor's Epilogue: Halina, 1948 to the Present', in Robinson, *Heaven, Hell, and Purgatory*, op cit., p. 171.
104. Boel Berner, 'Explaining Exclusion: Women and Swedish Engineering Education From the 1890s to the 1920s', *History and Technology*, 14 (1997): 26–27.
105. Duffin, 'Poisoning the Spindle (Part II)', p. 110.
106. Peter Twohig, *Labour in the Laboratory*. Montreal: McGill-Queen's University Press, 2005, 132.
107. 'Rona Hatt – CHBE Hall of Fame', CHBE Hall of Fame, https://chbe.ubc.ca/hall-of-fame/2021/rona-hatt. Accessed 28 May 2024.
108. Dominion Bureau of Statistics. *Census of Canada, 1951 – Occupation and Industry Trends in Canada*.
109. Duffin, 'Poisoning the Spindle (Part II)', p. 110.
110. Halina M. Robinson, 'Vinca revisited – another happenstance in the discovery of vinblastine', *Biochemistry and Cell Biology* 69 (1991): 581–582, p. 581.
111. Jacalyn Duffin, 'Appendix: The Story of Halina's Scientific Discovery', in Robinson, *Heaven, Hell, and Purgatory*, p. 177.
112. Duffin, 'Poisoning the Spindle (Part II)', p. 110.

113. Duffin, Editor's Prologue, pp. 172–173; 'Robert Noble, MD PhD', *Canadian Medical Hall of Fame*, www.cdnmedhall.ca/laureates/robertnoble. Accessed 20 May 2024.
114. Duffin, 'Poisoning the Spindle (Part II)', p. 110.
115. Duffin, 'Poisoning the Spindle (Part II)', p. 111. Which year is not known. Thanks to Jackie Duffin, Shelley McKellar, and Anne Quirk for helping try to pin down the year.
116. Duffin, 'Poisoning the Spindle (Part I)', p. 64.
117. Duffin, 'Poisoning the Spindle (Part II)', p. 110.
118. Halina Robinson, 'Vinca Revisited – Another Happenstance in the Discovery of Vinblastine', *Biochemistry and Cell Biology* 69 (1991), 581–582; Robert L. Noble, 'The Discovery of the Vinca Alkaloids-Chemotherapeutic Agents against Cancer', *Biochemistry and Cell Biology* 68 (1990), 1344–1351.
119. R. L. Noble, C. T. Beer, and J. H. Cutts, 'Role of Chance Observations in Chemotherapy: Vinca Rosea', *Annals of the New York Academy of Sciences*, 76 (1958): 882–894.
120. Duffin, 'Poisoning the Spindle (Part II)', p. 111.
121. Noble, 'The Discovery of the Vinca Alkaloids', op cit., p. 1345; Robinson, 'Vinca revisited', op cit., p. 582.
122. Duffin, 'Poisoning the Spindle (Part II)', p. 111.
123. 'Charles Beer, PhD', *Canadian Medical Hall of Fame*, www.cdnmedhall.ca/laureates/charlesbeer. Accessed 28 May 2024.
124. Robinson, 'Vinca revisited', p. 582.

Chapter 2

1. Reportedly said at a celebration for her 100th birthday, as reported by the Associated Press in, for example: *The Record* (Hackensack, New Jersey), 19 April 2009, p. A-3.
2. 'Wuppertal', Jewish Virtual Library. www.jewishvirtuallibrary.org/wuppertal. Accessed 15 September 2024.
3. Gertrude Shakno, Interviewed by Diane Plotkin, Dallas, Texas, 23 June 1991, Quoted in 'The Memoirs of Ellen Loeb', op cit., pp. 1–2. Gertrude's nickname takes on different spellings: Trudi and Trudy. See also Oral History Interview with Gertrude Shakno, Dallas Holocaust and Human Rights Museum, 20 November 2003, https://dhhrm.catalogaccess.com/archives/18458; Oral History with Gertrude Shakno, Dallas Holocaust and Human Rights Museum, 21 May 1991, https://dhhrm.catalogaccess.com/archives/10595.
4. Ibid.
5. 'Wuppertal', Jewish Virtual Library.
6. Shakno in 'The Memoirs of Ellen Loeb', pp. 3–4.
7. Pick, USC interview, op cit., tape 1.
8. Ibid., tape 2.
9. Mark Mazower, *Hitler's Empire: Nazi Rule in Occupied Europe*, London: Penguin Books, 2009, p. 55.
10. Wolf Gruner, Translated by Alex Skinner, *The Holocaust in Bohemia and Moravia*, New York & Oxford: Berghahn, 2019, pp. 68–69.
11. Mazower, *Hitler's Empire*, p. 55.
12. Pick, USC interview, tapes 1 and 2.
13. The city is now Berezne, Ukraine.
14. Robinson, *Heaven, Hell, and Purgatory*, op cit., p. 8.

15. Ibid., p. 24.
16. Berezne-guidebook, shtetl routes, https://shtetlroutes.eu/en/berezne-putvnik. Accessed 21 September 2024.
17. Robinson, *Heaven, Hell, and Purgatory*, p. 13.
18. Mazower, *Hitler's Empire*, p. 64.
19. Robinson, *Heaven, Hell, and Purgatory*, pp. 36–40.
20. See, e.g. Mordechai V. Bernstein, ed. *Berger, Jacob Solomon. The Zamosc Memorial Book* (Zamość, Poland), Pinkas Zamosc: New Jersey, 2004 [1957], available at www.jewishgen.org/yizkor/zamosc1/Zamosc1.html. Accessed 4 May 2025.
21. Robinson, *Heaven, Hell, and Purgatory*, pp. 65.
22. 'Warsaw', Holocaust Encyclopedia, United States Holocaust Memorial Museum, https://encyclopedia.ushmm.org/content/en/article/warsaw. Accessed 22 September 2024.
23. Neujahr, USC interview, op cit., tape 1.
24. Yisrael Gutman, *The Jews of Warsaw: 1939–1943: Ghetto, Underground, Revolt.* Bloomington and Indianapolis: Indiana University Press, 1989, p. xvi. For more on the nuanced views and actions of non-Jewish Poles, see, e.g. Connelly, John. 'Poles and Jews in the Second World War: The Revisions of Jan T. Gross', *Contemporary European History* 11, 4 (2002): 641–658.
25. Neujahr, USC interview, tape 1. Halina Neujahr, 'Testimony recorded in Jerusalem, 12–23 June 1981, for deposition in Yad Vashem', Halina Neujahrs Arkiv (ACC.NR. 2007/034), Arkiv nr 6754:1, F 1 a, v. 1. Unfortunately, I have been unable to find any further details about the circumstances of her father's death.
26. Neujahr, USC interview, tape 1.
27. Ibid.
28. Mazower, *Hitler's Empire*, p. 66.
29. Gutman, *The Jews of Warsaw*, op cit., p. 11. Mazower, *Hitler's Empire*, p. 71.
30. Mazower, *Hitler's Empire*, p. 74.
31. Gutman, *The Jews of Warsaw*, p. 11.
32. See, e.g. 'Italian Racial Laws', Italy and the Holocaust Foundation, www.italyandtheholocaust.org/italian-racial-laws.aspx#. Accessed 4 October 2024. Mazower, *Hitler's Empire*, p. 393.
33. Rita was born with the surname Levi, which was her father's surname. It was not until later in her career that she added Montalcini, her mother's maiden name, to her surname, resulting in Levi-Montalcini. See, e.g. Ruby Rohrlich, 'Jewish Lives: Rita Levi-Montalcini', *Judaism: A Quarterly Journal of Jewish Life and Thought* 49, 1 (2000): 36–55, pp. 37–38.
34. Levi-Montalcini, *In Praise of Imperfection*, p. 126.
35. Ibid., p. 84, 126–128. Marco Piccolino, 'Youthfulness, Vocations, and First Experiments of Rita Levi-Montalcini, scientist. First part: the difficult choice of a graduate in medicine', *Confinia Cephalalgica et Neurolgica* 31, 1 (2021): 1–7, pp. 4–5.
36. Levi-Montalcini, *In Praise of Imperfection*, p. 85.
37. Ibid., p. 88.
38. 'Jewish Population of Europe in 1933: Population Data by Country', United States Holocaust Memorial Museum, https://encyclopedia.ushmm.org/content/en/article/jewish-population-of-europe-in-1933-population-data-by-country. Accessed 28 September 2024.

39. Michael K. Silber, 'Budapest', The YIVO Encyclopedia of Jews in Eastern Europe, https://encyclopedia.yivo.org/article.aspx/Budapest. Accessed 28 September 2024.
40. 'Historical Background: The Jews of Hungary During the Holocaust', Yad Vashem, www.yadvashem.org/articles/general/jews-of-hungary-during-the-holocaust.html. Accessed 28 September 2024.
41. See, e.g. M. Mária Kovács, 'Disenfranchised by Law – The "Numerus Clausus" in Hungary 1920–1945', *Shoah: Intervention, Methods. Documentation* 1, 2 (2014): 136–143.
42. Klein, USC interview, op cit., tape 1.
43. 'Museum Condemns Attempts to Rehabilitate Hungarian Fascist Leader', United States Holocaust Memorial Museum, Press Release dated 28 June 2017, www.ushmm.org/information/press/press-releases/museum-condemns-attempts-to-rehabilitate-hungarian-fascist-leader. Accessed 28 September 2024.
44. See, e.g. 'Historical Background: The Jews of Hungary During the Holocaust', op cit.
45. Klein, USC interview, tapes 1 and 2.
46. Ibid., tape 1.
47. See, e.g. Mazower, *Hitler's Empire*, pp. 397–400.
48. Robinson, *Heaven, Hell, and Purgatory*, pp. 27–28.
49. Neujahr, USC interview, tape 1.
50. Mazower, *Hitler's Empire*, pp. 89–96.
51. Robinson, *Heaven, Hell, and Purgatory*, p. 61.
52. Ibid., p. 74.
53. Anna Krasińska, 'Record of Witness Testimony 370, The Polish Research Institute in Lund, 19 June 1946, p. 2. Witness testimony in the collection of the Polish Research Institute in Lund, Sweden, Lund University Library. Bracketed information in original source.
54. Neujahr, USC interview, tape 2.
55. Gutman, *The Jews of Warsaw*, pp. 62–63.
56. Ibid., pp. 53–55.
57. Ibid., p. 63.
58. Ibid., pp. 104–105.
59. Ludwick Hirszfeld, *The Story of One Life*. eds. Marta A. Balińska and William H. Schneider, trans. Marta A. Balińska, Rochester, NY: University of Rochester Press, 2013, p. 201. See also Charles G. Roland, *Courage Under Siege: Starvation, Disease, and Death in the Warsaw Ghetto*. New York: Oxford University Press, 1992.
60. Hirszfeld, *The Story of One Life*, p. 202.
61. Neujahr, USC testimony, tape 2.
62. This was an acknowledged part of her history already in the early 1970s. See, e.g. Adamczyk and Carlösten, 'Hon började sin forskarbana I gettot', op cit., p. 17.
63. 'Anti-Jewish policy after the establishment of the Protectorate of Bohemia and Moravia', Terezín Initiative Institute, www.holocaust.cz/en/history/final-solution/the-final-solution-of-the-jewish-question-in-the-bohemian-lands/anti-jewish-policy-after-the-establishment-of-the-protectorate-of-bohemia-and-moravia. Accessed 19 October 2024.
64. Ruth Pick, USC interview, tape 2. See also, Gruner, *The Holocaust in Bohemia and Moravia*, op cit., pp. 84–92.
65. Gruner, *The Holocaust in Bohemia and Moravia*, pp. 42–43; Rob Cameron, 'Czech doctors' association to apologise for 1938 anti-Jewish purge', Radio Prague, https://

english.radio.cz/czech-doctors-association-apologise-1938-anti-jewish-purge-8575635, last accessed 19 October 2024.

66. Gruner, *The Holocaust in Bohemia and Moravia*, pp. 92–93.
67. Ruth Bondy, *Trapped: Essays on the History of the Czech Jews, 1939–1943*. Jerusalem: Yad Vashem, 2008, p. 197.
68. Pick, USC interview, tape 2.
69. Gruner, *The Holocaust in Bohemia and Moravia*, pp. 268–273.
70. Ibid., p. 294.
71. Ibid., p. 223.
72. Pick, USC interview, tape 2.
73. Ibid.
74. Levi-Montalcini, *In Praise of Imperfection*, p. 89.
75. Ibid., p. 90.
76. Ibid., p. 91.
77. Ibid., pp. 91–92.
78. Ibid.
79. Klein, USC interview, tape 2.
80. Trudy Shakno, Oral History Interview 2003, Dallas Holocaust and Human Rights Museum (DHHRM), part 1.
81. 'The Memoirs of Ellen Loeb', op cit., pp. 21–22.
82. Shakno in 'The Memoirs of Ellen Loeb', p. 3.
83. 'Julius Loeb', Joods Monument, www.joodsmonument.nl/en/page/217126/julius-loeb, last accessed 18 October 2024.
84. Trudy Shakno, DHHRM Oral History Interview, 1990, part 1.
85. Trudy Shakno, Oral History Interviews, Dallas Holocaust and Human Rights Museum, 1990, 1991 and 2003. On the Jewish hospitals in Amsterdam, see: J. Y. Vanderhoek, 'Jewish Hospitals in 20th Century Amsterdam: A Tale of Growth, Change, and Decline', *Rambam Maimonides Medical Journey* 14, 4 (2023).
86. Shakno in 'The Memoirs of Ellen Loeb', p. 4.
87. 'Amsterdam', Holocaust Encyclopedia, United States Holocaust Memorial Museum, https://encyclopedia.ushmm.org/content/en/article/amsterdam. Accessed 18 October 2024.

Chapter 3

1. Loeb, 'The Memoirs of Ellen Loeb', p. 14.
2. Quoted in Mazower, *Hitler's Empire*, p. 398.
3. Hannah Arendt, *Eichmann in Jerusalem: A Report on the Banality of Evil*, New York: Penguin Books, 2006 [1963], p. 114.
4. See 'Camp Westerbork', www.kampwesterbork.nl/en. Accessed 2 November 2024.
5. Loeb, 'The Memoirs of Ellen Loeb', p. 14
6. Ibid., pp. 15–16.
7. Ibid., pp. 17–18.
8. Ibid., pp. 18–19.
9. Ibid., p. 16.
10. Livia Rothkirchen, *The Jews of Bohemia and Moravia: Facing the Holocaust.* Lincoln, Nebraska and Jerusalem, Israel: University of Nebraska Press and Yad Vashem, 2005, p. 134. On Theresienstadt and the Fortress of Terezín, see 'The Fortress of Terezín',

UNESCO World Heritage Convention, https://whc.unesco.org/en/tentativelists/1561. Accessed 2 November 2024.
11. Rothkirchen, *The Jews of Bohemia and Moravia*, pp. 233–264.
12. 'Theresienstadt', Holocaust Encyclopedia, The United States Holocaust Memorial Museum, https://encyclopedia.ushmm.org/content/en/article/theresienstadt. Accessed 2 November 2024.
13. Pick, USC interview, tape 2. Arolsen Archives, Ghetto Theresienstadt Card File, DocID: 5070115 (ALFRED PICK), https://collections.arolsen-archives.org/en/document/5070115. Rothkirchen, *The Jews of Bohemia and Moravia*, p. 133.
14. Rothkirchen, *The Jews of Bohemia and Moravia*, pp. 127–128.
15. Pick, USC interview, tape 2.
16. Erich Springer, 'Health Conditions in Terezín', in *Terezín*, Prague: The Council of Jewish Communities in the Czech Lands, 1965, p. 199.
17. Margalit Shlain, 'Nursing in the Theresienstadt Ghetto', *NASHIM* 36 (2020): 60–85, p. 64.
18. Anna Hájková, 'Medicine in Theresienstadt', *Social History of Medicine* 33, 1 (2018): 79–105, pp. 79–80.
19. Shlain, 'Nursing in the Theresienstadt Ghetto', p. 65.
20. Pick, USC interview, tape 2.
21. Hájková, 'Medicine in Theresienstadt', p. 100.
22. Pick, USC interview, tape 2.
23. Springer, 'Health Conditions in Terezín', p. 119.
24. Loeb, 'The Memoirs of Ellen Loeb', pp. 21–22.
25. Springer, 'Health Conditions in Terezín', p. 121.
26. Pick, USC interview, tape 3.
27. Shlain, 'Nursing in the Theresienstadt Ghetto', p. 65.
28. 'The Memoirs of Ellen Loeb', p. 22. Arolsen Archives, Personal file of LÖB, DINA, born on 12-Jul-1905, 1874778 (DINA LÖB), https://collections.arolsen-archives.org/en/document/1874778.
29. Pick, USC interview, tape 3. Arolsen Archives, Ghetto Theresienstadt Card File, DocID: 5070115 (ALFRED PICK), https://collections.arolsen-archives.org/en/document/5070115.
30. Mazower, *Hitler's Empire*, pp. 303–312.
31. Gutman, *The Jews of Warsaw*, pp. 72–74. Mazower, *Hitler's Empire*, p. 311.
32. Neujahr, USC interview, tape 3.
33. United States Holocaust Memorial Museum (USHMM), 'Poles: Victims of the Nazis', 2000, p. 18.
34. Robinson, *Heaven, Hell, and Purgatory*, p. 80.
35. Gutman, *The Jews of Warsaw*, p. 197.
36. Neujahr, USC interview, tapes 2 and 3.
37. Robinson, *Heaven, Hell, and Purgatory*, pp. 83–84.
38. Gutman, *The Jews of Warsaw*, pp. 307–311.
39. Neujahr, Yad Vashem testimony, op cit., p. 6.
40. Neujahr, USC testimony, tape 4.
41. Robinson, *Heaven, Hell, and Purgatory*, p. 84.
42. Mazower, *Hitler's Empire*, p. 494.
43. USHMM, 'Poles: Victims of the Nazis', op cit., pp. 17–18.

44. Robinson, *Heaven, Hell, and Purgatory*, p. 82.
45. Mazower, *Hitler's Empire*, p. 494.
46. USHMM, 'Poles: Victims of the Nazis', p. 23.
47. Robinson, *Heaven, Hell, and Purgatory*, p. 102.
48. Ibid., p. 105. See also Arolsen Archives, A.E.F. DP Registration Card, DocID: 69506240 (Halina CZAJKOWSKA), https://collections.arolsen-archives.org/en/document/69506240.
49. Robinson, *Heaven, Hell, and Purgatory*, p. 106.
50. Ibid., pp. 106–110.
51. 'The unloading ramps and selections', Auschwitz-Birkenau Museum and Memorial, www.auschwitz.org/en/history/auschwitz-and-shoah/the-unloading-ramps-and-selections. Accessed 9 November 2024.
52. 'Camp admission process', State Museum at Majdanek, www.majdanek.eu/en/history/camp_admission_process/7. Accessed 9 November 2024.
53. Neujahr, Yad Vashem testimony, p. 8.
54. Ibid., p. 9. Neujahr, USC testimony, tapes 4 and 5.
55. Robinson, *Heaven, Hell, and Purgatory*, p. 110.
56. Ibid., pp. 110–111.
57. Ibid., pp. 111–115.
58. Loeb, 'The Memoirs of Ellen Loeb', p. 23.
59. Quoted in Loeb, 'The Memoirs of Ellen Loeb', p. 23.
60. Loeb, 'The Memoirs of Ellen Loeb', p. 24.
61. Ibid., pp. 23–25.
62. Pick, USC testimony, tape 3.
63. Mazower, *Hitler's Empire*, 310–311. 'Nazi Camps – Background Information', Forced Labor 1939–1945 Memory and History, www.zwangsarbeit-archiv.de/en/zwangsarbeit/erfahrungen/lager/index.html. Accessed 23 November 2024.
64. 'Satellite Camp Lenzing', Mauthausen Guides, www.mauthausen-guides.at/en/subcamp/satellite-camp-lenzing, accessed 22 November 2024. Personal File of Ilse Loeb, Mauthausen Concentration Camp, Arolsen Archives, DocID: 1874790 (ILSE LÖEB), https://collections.arolsen-archives.org/en/document/1874790. Accessed 22 November 2024.
65. Loeb, 'The Memoirs of Ellen Loeb', p. 26.
66. Ibid., p. 28.
67. 'Oederan Subcamp', KZ-Gedenkstätte Flossenbürg, www.gedenkstaette-flossenbuerg.de/en/history/satellite-camps/oederan, accessed 22 November 2024. Personal file of Pick, Ruth, Flossenbürg Concentration Camp, Arolsen Archive, DocID: 11081129 (RUTH PICK), https://collections.arolsen-archives.org/en/document/11081129. Accessed 22 November 2024.
68. Pick, USC Testimony, tapes 3 and 4.
69. Neujahr, Yad Vashem Testimony, p. 10.
70. Ibid.
71. On the camp, see: Felicja Karay. Sara Kitai, trans. *Death Comes in Yellow: Skarzysko-Kamienna Slave Labor Camp*. Taylor & Francis Group, 1997. The statistic of 70 per cent comes from: Christopher R. Browning, 'Death Comes in Yellow: Skarzysko-Kamienna Slave Labor Camp Felicja Karay (Book Review)', *Holocaust and Genocide Studies*, 1 (2018): 475–477, pp. 475–476.

72. Neujahr, Yad Vashem Testimony, pp. 10–11.
73. Robinson, *Heaven, Hell, and Purgatory*, p. 125.
74. Ibid., p. 130.
75. Ibid., pp. 131–132.
76. 'Nazi-era Forced Labor: A Commonplace Mass Crime', nsdoku münchen, www.nsdoku.de/en/neuaubing-memorial-site/nazi-era-forced-labor. Accessed 23 November 2024.
77. Levi-Montalcini, *In Praise of Imperfection*, pp. 91–93.
78. Quoted in Liversidge, 'Interview: Rita Levi-Montalcini', p. 73. Brackets in original.
79. The discovery of the organiser effect was made in collaboration with a female embryologist named Hilde Mangold, who was Spemann's doctoral student at the same time as Viktor Hamburger in the early 1920s. She died in 1924 and the credit for the discovery fell to Spemann. See Ellen Elliott, 'Women in Science: Hilde Mangold and the embryonic organizer', *The Jackson Laboratory*, JAX Blog, Oct. 20, 2016, www.jax.org/news-and-insights/jax-blog/2016/october/women-in-science-hilde-mangold.
80. Quoted in Liversidge, 'Interview: Rita Levi-Montalcini', p. 74. Italics in original.
81. Neither Viktor Hamburger, Spemann's former student, nor Giuseppi Levi – both of whom were Jewish – appear to have been concerned about Spemann or his alleged Nazi sympathies either. See Rita Levi-Montalcini, 'One of Hans Spemann's Pupils', In *Studies in Developmental Neurobiology: Essays in Honor of Viktor Hamburger*, edited by W. Maxwell Cowan. New York: Oxford University Press, 1981. On Spemann as an alleged Nazi sympathiser, see for example: C. Brandt, *Vitalism, Holism, and Metaphorical Dynamics of Hans Spemann's 'Organizer' in the Interwar Period.* J Hist Biol 55 (2002): 285–320. 'Medicine: Nobel Prize', *Time*, November 4, 1935, https://time.com/archive/6754517/medicine-nobel-prize-5. Accessed 23 November 2024.
82. Levi-Montalcini, 'One of Hans Spemann's Pupils', p. 34.
83. Levi-Montalcini, *In Praise of Imperfection*, p. 94.
84. Piergiorgio Strata, 'Rita Levi-Montalcini and her major contribution to neurobiology', *Rendiconti Lincei. Scienze Fisiche e Naturali.* 29 (2018): 737–753, p. 746.
85. Levi-Montalcini, *In Praise of Imperfection*, p. 95.
86. Ibid., pp. 101–104.
87. Ibid., pp. 104–105.
88. See e.g. Peter Curry, 'The Detonation of the Bridges of Florence and German Atrocities in Wartime Italy During World War Two', History Hit, October 29, 2018, www.historyhit.com/the-detonation-of-the-bridges-of-florence-and-german-atrocities-in-wartime-italy-during-world-war-two; Chiara Ulivi, 'The Florentine artistic heritage and the Second World War', Le Gallerie Degli Uffizi, 1 October 2024, www.uffizi.it/en/news/il-patrimonio-artistico-fiorentino-e-la-seconda-guerra-mondiale.
89. Levi-Montalcini, *In Praise of Imperfection*, p. 105.
90. Ibid., p. 105.
91. Italy, Holocaust Encyclopedia, USHMM, https://encyclopedia.ushmm.org/content/en/article/italy. Accessed 24 November 2024.
92. Mazower, *Hitler's Empire*, pp. 396–400.
93. Klein, USC interview, tape 2.
94. David S. Wyman, *The Abandonment of the Jews: America and the Holocaust 1941–1945* (New York: Pantheon Books, 1984), p. 235.
95. 'Deportation of Hungarian Jews Begins, May 15, 1944', History Unfolded, USHMM, https://newspapers.ushmm.org/events/deportation-of-hungarian-jews-begins, accessed 8 December 2024.

96. Randolph L. Braham, 'The Holocaust in Hungary: A Retrospective Analysis', in *The Nazis' Last Victims: The Holocaust in Hungary*, edited by Rodolph L. Braham and Scott Miller. Detroit: Wayne State University Press, 1998, p. 37.
97. Klein, USC interview, tape 3.
98. László Karsai, 'The Last Phase of the Hungarian Holocaust: The Szálasi Regime and the Jews', in *The Nazis' Last Victims: The Holocaust in Hungary*, Randolph I. Braham and Scott Miller, eds. Detroit: Wayne State University Press, 1997, pp. 103–116.
99. Ibid., p. 112.
100. 'Historical Background: The Jews of Hungary During the Holocaust', Yad Vashem, www.yadvashem.org/articles/general/jews-of-hungary-during-the-holocaust.html. Accessed 12 December 2024.
101. Klein, USC interview, tape 3.
102. Ibid.
103. Victor Hugo, Lascelles Wraxall, trans. *Les Misérables*. Norwalk, Connecticut: The Easton Press, 1966 [1862], p. 95.
104. See, e.g. Daina S. Eglitis, 'Silences of Memory: Liberator Sexual Assault in the East at the End of World War II', *The Journal of Holocaust Research*, 38, 3–4 (2024): 316–335. Mary Louise Roberts, *What Soldiers Do: Sex and the American GI in World War II France*. The University of Chicago Press, 2013.
105. Levi-Montalcini, *In Praise of Imperfection*, p. 107.
106. James Mark. 'Remembering Rape: Divided Social Memory and the Red Army in Hungary 1944–1945', *Past & Present* 188, 1 (2005): 133–161. Krisztián Ungváry, Ladislaus Lob, trans. *Battle for Budapest: 100 Days in World War II*. New York: I.B. Tauris, 2005. Péter Sasvári, 'The Siege of Budapest – A Terrible Winter', *Hungarian Conservative*, 30 December 2023, www.hungarianconservative.com/articles/culture_society/siege-of-budapest_wwii_winter-1944-1945_germans_hungarians_red-army.
107. Klein, USC interview, tape 4.
108. Robinson, *Heaven, Hell, and Purgatory*, p. 134.
109. 'The Concentration Camp (1943–1945)', Gedenkstätte Bergen-Belsen, https://bergen-belsen.stiftung-ng.de/en/history/theconcentrationcamp1943-1945. Accessed 12 January 2025.
110. Pick, USC interview, tape 4. Rothkirchen, *The Jews of Bohemia and Moravia*, p. 263. 'Oederan Subcamp', KZ-Gedenkstätte Flossenbürg, www.gedenkstaette-flossenbuerg.de/en/history/satellite-camps/oederan. Accessed 22 November 2024. 'Theresiendstadt: Final Weeks, Liberation, and Postwar Trials', USHMM, https://encyclopedia.ushmm.org/content/en/article/theresienstadt-final-weeks-liberation-and-postwar-trials. Accessed 12 January 2025.
111. Loeb, 'The Memoirs of Ellen Loeb', pp. 37–38. See also 'Satellite Camp Lenzing', Mauthausen Guides', www.mauthausen-guides.at/en/subcamp/satellite-camp-lenzing. Accessed 12 January 2025.

Chapter 4

1. Robinson, *Heaven, Hell, and Purgatory*, p. 159.
2. Neujahr, USC interview, tape 6.
3. William Arthur Wood, Oral history, reel 3. Imperial War Museum. www.iwm.org.uk/collections/item/object/80014962.
4. Robinson, *Heaven, Hell, and Purgatory*, p. 136.

5. Deborah E.-S. Hemstreet and George M. Weisz, 'One Page in the History of Starvation and Refeeding', *Rambam Maimonides Medical Journal* 15, 2 (2024).
6. Robinson, *Heaven, Hell, and Purgatory*, p. 140.
7. Wood, Oral history, reel 4.
8. Laurence Geoffrey Rowland Wand, Oral History, reel 2. Imperial War Museum. www.iwm.org.uk/collections/item/object/80008874.
9. Robinson, *Heaven, Hell, and Purgatory*, p. 144.
10. Neujahr, Yad Vashem testimony, pp. 9–10.
11. Neujahr, USC interview, tape 7.
12. Ibid. On the disinfection process, see Roman Wasserman Wroblewski, *The Liberated 1945: White Boat Mission from Bergen-Belsen to Sweden*. Stockholm: GML Förlag AB, 2021, p. 69.
13. Robinson, *Heaven, Hell, and Purgatory*, pp. 147–148.
14. Victoria Van Orden Martínez and Malin Thor Tureby, 'Monuments Cast Shadows: Remembering and Forgetting the "Dead Survivors" of Nazi Persecution in Swedish Cemeteries', in *Fallen Monuments and Contested Memorials*, ed. Juilee Decker. London: Routledge, 2023, 177–189; Wroblewski, *The Liberated 1945*.
15. Robinson, *Heaven, Hell, and Purgatory*, p. 139.
16. Loeb, 'The Memoirs of Ellen Loeb', pp. 38–39.
17. Letter from the Strauss family to the Shakno family, 16 August 1945, cited in Loeb, 'The Memoirs of Ellen Loeb', p. 43.
18. Dina Loeb to the Shaknos, 18 September 1945, cited in Loeb, 'The Memoirs of Ellen Loeb', pp. 45–46.
19. Evelien Gans, '"The Jew" as Dubious Victim', In: R. Ensel and E. Gans, eds. *Holocaust, Israel and 'the Jew': Histories of Antisemitism in Postwar Dutch Society*. Amsterdam University Press, 2016, pp. 61–82.
20. Pick, USC interview, tape 4.
21. Jan Láníček, 'After the Whirlwind: Jewish Absence in Postwar Czechoslovakia', *Journal of Contemporary History* 52, 2 (2017): 278–96, p. 280.
22. e.g. 'Czech Republic Historical Background', Yad Vashem, www.yadvashem.org/righteous/stories/czech-republic.html. Accessed 23 August 2025.
23. A. Hájková, 'To Terezín and Back Again: Czech Jews and their Bonds of Belonging from Deportations to the Postwar', *Dapim: Studies on the Holocaust* 28, 1 (2014): 38–55, p. 50.
24. Ruth Pick, USC interview, tape 4.
25. Cited in Ungváry, *Battle for Budapest*, op cit., p. 283.
26. Pablo Gorondi, 'Hungary honors its 700,000 victims of Soviet labor camps', The Associated Press, 13 May 2015, https://apnews.com/general-news-eaa56bb88684410d987da53d255c7367. Accessed 23 August 2025.
27. Klein, USC interview, tape 4.
28. Ungváry, *Battle for Budapest*, p. 283.
29. Ibid., pp. 216–310.
30. Eva Klein, USC interview, tape 5.
31. Levi-Montalcini, *In Praise of Imperfection*, pp. 34–39.
32. Ibid., p. 107.
33. Ibid., pp. 106–108.
34. Ibid., p. 111.

35. Ellen to the Shaknos, 25 July 1945, cited in Loeb, 'The Memoirs of Ellen Loeb', p. 42.
36. Ellen to Trudi, 15 July 1945, cited in Loeb, 'The Memoirs of Ellen Loeb', p. 41.
37. An excellent master's thesis on the topic is Dorien van Duivenboden, 'Migrant Nurses in the Netherlands 1945–1975', MA thesis, Leiden University, 2019.
38. Ellen to Trudi, 24 August 1945, cited in Loeb, 'The Memoirs of Ellen Loeb', p. 44.
39. Ellen to Trudi, 15 July 1945, cited in Loeb, 'The Memoirs of Ellen Loeb', p. 41.
40. Trudi to Ellen and Dina, undated but probably May 1945, cited in in Loeb, 'The Memoirs of Ellen Loeb', p. 40.
41. Shakno, Oral History Interview, DHHRM, 1991. A 1980 article in the *Texas Jewish Post* claims she studied at the University of Amsterdam (Jess Jawin, 'Solidarity Day', *Texas Jewish Post*, xxxiv, 18 (1 May 1980), p. 1.) I have contacted the university to try to verify this but have received no response.
42. Eva Klein, USC interview, tape 5.
43. Eva Klein, 'Self Portrait', Karolinska Institutet via Web Archive, https://web.archive.org/web/20170107050115/http://ki.se/en/mtc/eva-klein-group. Accessed 23 August 2025.
44. Ibid.
45. Eva Klein, USC interview, tape 5.
46. László Borhi, *Hungary in the Cold War: Between the United States and the Soviet Union*, Budapest: Central European University Press, 2004.
47. Robinson, *Heaven, Hell and Purgatory*, p. 158.
48. Ibid., p. 94.
49. Ibid., pp. 154–155.
50. Ibid., pp. 158–159. See also Martínez, 'History dies deep in the woods'.
51. Neujahr, Yad Vashem testimony, p. 12.
52. Archive of The Polish Research Institute in Lund, Volume 32:1:11, 'Oczy swiata' (The world's eyes) by Halina Neujahr, dated July 1945, Lund University Library, Sweden.
53. Neujahr, Yad Vashem testimony, p. 12.
54. Victoria Van Orden Martínez, *Afterlives: Jewish and Non-Jewish Polish Survivors of Nazi Persecution in Sweden Documenting Nazi Atrocities, 1945–1946* (doctoral dissertation, Linköping University Press, 2023), pp. 190–192.
55. Neujahr, USC interview, tape 7.
56. In her USC Shoah Foundation interview, Ruth states incorrectly that it was the 1932 census.
57. Pick, USC interview, tape 4.
58. Ibid.
59. Victoria Van Orden Martínez, 'Shaping Ongoing Survival in a Swedish Refugee Camp: A refugee-centered history of Jewish and non-Jewish survivors of Nazi persecution in Sweden', *Nordisk Judaistik – Scandinavian Jewish Studies*, 33, 1 (2002): 19–36.
60. *Bilder från Öreryd* (Öreryd: Öreryds hembygdsföreningen, 1990), 200–201.
61. Neujahr, USC interview, tape 7.
62. Robinson, *Heaven, Hell and Purgatory*, p. 160.
63. Ibid., pp. 159–160.
64. Halina Neujahrs Arkiv, Nordiska Museet, Stockholm, 6754:1/K 1.
65. Robinson, *Heaven, Hell and Purgatory*, p. 161.
66. Wyman, *The Abandonment of the Jews*, op cit., p. 340.

67. See, e.g. 'Immigration in Post-War America', Harry S. Truman Library & Museum, www.trumanlibrary.gov/education/presidential-inquiries/immigration-post-war-america. 'Post-War Years', U.S. Citizenship and Immigration Services, www.uscis.gov/about-us/our-history/explore-agency-history/overview-of-agency-history/post-war-years. 'December 22, 1945: New Directive on Immigrant Visas to the US', United States Holocaust Memorial and Museum, https://encyclopedia.ushmm.org/content/en/timeline-event/holocaust/1942-1945/truman-directive-on-immigrant-visas. Accessed 30 August 2025.
68. Affidavit dated 4 September 1945, and signed by Carl M. Loeb, The Dr. Ellen Loeb-Katz Collection, Dallas Holocaust and Human Rights Museum (DHHRM). See also Shakno, Oral History Interview 2003, DHHRM, part 1.
69. Levi-Montalcini, *In Praise of Imperfection*, p. 94.
70. Ibid., p. 113.
71. Ibid., pp. 113–117.
72. Georg Klein and Eva Klein, 'How One Thing Has Led to Another', *Annual Review of Immunology* 7 (1989): 1–33, p. 7.
73. Ibid., p. 5.
74. Ibid., pp. 4–5.
75. Ibid., p. 7.
76. Ibid., pp. 7–8.
77. Ibid., p. 9.
78. Gordon, *All Our Lives*, pp. 119–120.
79. O.N. 'Czechoslovakia Under Communist Rule', *The World Today*, 6, 1 (Jan. 1950): 15–27, 21. The journal was published by Chatham House (The Royal Institute of International Affairs), a British thinktank established in 1920 and is still operational today.
80. Pick, USC interview, tape 4.

Chapter 5

1. Eva Klein and Georg Klein, 'Reflections on Denmark and Hungary During the Second World War', Scandinavian Jewish Forum, January 2014. https://scandinavianjewish.blogspot.com/2014/01/eva-georg-klein.html. Accessed 1 March 2025. (Note in cited reference: 'Previously written and published 1993 DIMENSIONS.')
2. e.g. Sune Persson, 'Folke Bernadotte and the White Buses', in *'Bystanders' to the Holocaust: A Re-evaluation*, David Cesarini and Paul A. Levine, eds. Routledge, 2002.
3. e.g. Per Tistad, 2015. 'Repatriering till Polen 1945', *Flottans Män* 1 (2015): 30–1.
4. Andrzej Nils Uggla. *I nordlig hamn: polacker i Sverige under andra världskriget*, trans. Lennart Ilke. Uppsala multiethnic papers: 40. Uppsala: Centrum för multietnisk forskning, Uppsala Universitet, 1997, pp. 225–227.
5. Robinson, *Heaven, Hell, and Purgatory*, p. 162.
6. Memorandum 'Sak nr. 160 6/3 1946', F1 AC-4815 Emelie Czajkowska, Rödakorsflykting, Statens utrymningskommission Kanslibyrån (SUK), Riksarkivet, Stockholm, Sweden.
7. F1 AC-4815 Emelie Czajkowska, SUK.
8. Ibid.
9. Studentexamen 1947, Halina Neujahrs Arkiv, F1 a 2 (Nordiska Museet).
10. Robinson, *Heaven, Hell and Purgatory*, p. 163.

11. The Warsaw Institute provides a compelling description and photos of the destruction of Warsaw. See Katarzyna Utracka, 'Warsaw – the City that is No More', https://warsawinstitute.org/warsaw-city-no. Accessed 6 September 2025.
12. Victoria Martínez, 'The unusual way Sweden once solved its housing crisis and boosted living conditions', *The Local Sweden*, 17 May 2019, www.thelocal.se/20190517/the-unusual-way-sweden-once-solved-its-housing-crisis-and-improved-living-conditions.
13. Robinson, *Heaven, Hell and Purgatory*, pp. 166–167.
14. F1 B-1728 Halina Neujahr, Rödakorsflykting, Statens utrymningskommission Kanslibyrån (SUK), Riksarkivet (RA), Stockholm, Sweden.
15. E. Walter Holmstedt, 'Några upplysningar om STI:s versamhet', *STI-Tidningen* 5, 1 (1 February 1949), p. 8.
16. Företeckning över utländska elever vid S.T.I. Höstterminen 1947. Stockholms Tekniska Institut (STI) Övriga elevförteckningar, Stockholms stadsarkiv, Kungsklippan, Sweden.
17. 'Förteckning över tekniska läroanstalter', *STI-Tidningen* 5, 1 (1 February 1949), p. 7.
18. Holmstedt, 'Några upplysningarp', op cit., p. 8.
19. Robinson, *Heaven, Hell and Purgatory*, p. 167.
20. Student record for Czajkowska, Halina, Stockholms Tekniska Institut (STI) Betygskopier, utexaminderade elever, Stockholms stadsarkiv, Kungsklippan, Sweden.
21. Robinson, *Heaven, Hell and Purgatory*, p. 167.
22. Ibid., p. 167.
23. Friedman, 'The Road Back for the DP's', op cit.
24. Duffin, 'Editor's Epilogue' in Robinson, *Heaven, Hell and Purgatory*, p. 171.
25. Student record for Czjkowska, Halina, Stockholms Tekniska Institut (STI) Betygskopier, utexaminderade elever, Stockholms stadsarkiv, Kungsklippan, Sweden.
26. Vårterminen 1950, Matrikel över utexaminderade elevet från Stockholms Tekn. Institutet, HT1947-HT59, Stockholms Tekniska Institut (STI) Elevmatriklar och -liggare, Stockholms stadsarkiv, Kungsklippan, Sweden.
27. 'Nya STI-ingenjörer', STI-Tidningen 6, 3 (September 1950), p. 12.
28. Neujahr, USC interview, tape 7.
29. Ibid.
30. Rydberg, 'Forskare mot all odds', op cit., p. 9. Translated from Swedish.
31. Neujahr, USC interview, tape 7. F1 B-1728 Halina Neujahr, SUK, RA.
32. SOU 1995:10, pp. 222–223.
33. Neujahr, USC interview, tape 8.
34. F1 B-1728 Halina Neujahr, SUK, RA.
35. Letter from Halina Neujahr to the Swedish Justice Department, dated 30 June 1953, F1 B-1728 Halina Neujahr, SUK, RA.
36. F1 B-1728 Halina Neujahr, SUK, RA.
37. Bodin, Anna. 'Vi möttes på en bräda', Dagens Nyheter (Stockholm, Sweden), 28 July 2015, p. 19.
38. Klein, USC interview, tape 5.
39. 'Timeline: KI through the centuries', Karolinska Institutet, https://ki.se/en/about-ki/history-and-cultural-heritage/medical-history-and-heritage/the-history-of-karolinska-institutet/timeline-ki-through-the-centuries, accessed 5 April 2025.
40. Alina M. Lindegren, *Education in Sweden*. Bulletin 1952, No. 17. Washington, DC: United States Government Printing Office, 1952, p. 44.

41. See, e.g. László Borhi, *Hungary in the Cold War, 1945–1956: Between the United States and the Soviet Union*. Budapest: Central European University Press, 2004.
42. László Varga, Trans. Anna Szalai. 'Rákosi, Mátyás', The YIVO Encyclopedia of Jews in Eastern Europe, https://encyclopedia.yivo.org/article/1626, accessed 5 April 2025. See also R. G. Szabó (2024). 'Anti-Zionism in the show trials in Rákosi era Hungary (1948–1953)', *Journal of Modern Jewish Studies* 1 (2024): 1–26.
43. Klein and Klein, 'How one thing has led to another', op cit., p. 7.
44. Valery N. Soyfer, 'The consequences of political dictatorship for Russian science', *Nature Reviews Genetics* 2, 9 (2001): 723–729, p. 725.
45. H. J. Muller, 'Genetics in the Scheme of Things', in *Proceedings of the Eighth International Congress of Genetics*, edited by Gert Bonnier and Robert Larsson, Lund: Berlingska Boktrykeriet, 1949, p. 106. See also Vadim J. Birstein, *The Perversion of Knowledge: The True Story of Soviet Science*. Boulder, Colo.: Westview Press, 2001.
46. Klein, USC interview, tape 5.
47. Magdolna Hargittai, *Women Scientists: Reflections, Challenges, and Breaking Boundaries*. Oxford University Press, 2015, p. 46.
48. George J. Race, ed. *UT Southwestern: Commemorating the First Half-Century*, Student and Alumni Affairs Office of The University of Southwestern Medical Foundation, 1998, pp. 1–3. UT Southwestern Medical School Collection (UTS), MSS0016. UT Southwestern Archives & Special Collections.
49. Catharine Schulze, *What Happened During 1943–1968: The First Twenty-Five Years at Southwestern Medical School in Dallas*, Dallas, Texas, 1971, p. 60. UTS, MSS0016, Box 1, Folder 13.
50. Certified letter dated 4 January 1946, signed by Ilse (Ellen) Loeb, The Dr. Ellen Loeb-Katz Collection, 2002.34.42, Dallas Holocaust and Human Rights Museum (DHHRM).
51. Also spelled *Joodse Invalide*. Kroonenberg was also in Theresienstadt. On Kroonenberg, see The Felix Levenbach & Irma Levenbach née Kroonenberg collection, Ghetto Fighters House Archives, www.infocenters.co.il/gfh/notebook_ext.asp?item=58242&site=gfh&lang=ENG&menu=1.
52. Letter of recommendation for Ellen written by Dr S. Kroonenberg, 7 January 1946, Loeb-Katz Collection, 2002.34.43, DHHRM. Translation from Dutch.
53. Bulletin of Southwestern Medical College, Catalogue Number, Volume V, Dallas, Texas, August 1948, pp. 56–57. UTS, MSS0016.
54. Shakno, Oral History Interview, 1991, DHHRM. Trudy says Ellen attended Southern Methodist University (SMU), where she earned a Bachelor's degree; however, there is no record of Ellen attending SMU. Moreover, the timing given by Trudy for Ellen's period of study at SMU conflicts with her time at Southwestern.
55. Race, *UT Southwestern*, op cit., p. 11.
56. Shakno, DHHRM Oral History Interview, 1991.
57. *Caduceus*, Southwestern Medical College, 1949. UTS, MSS0016.
58. Kenneth D. Roseman, 'Six-Tenths of a Percent of Texas', Chapter 15 in *Lone Stars of David*, 2007, p. 207.
59. Shakno, DHHRM Oral History Interview, 2003, part 1.
60. *Caduceus*, 1949, op cit.
61. Shakno, DHHRM Oral History Interview, 1991.
62. Bulletin of Southwestern Medical College, op cit., pp. 65–66.

63. Commencement Program and Speech, 1952, 2, UTS, MSS0016, Box: 13, Folder: 7; Southwestern Medical School class of 1952 composite photo, 1952, UT Southwestern Digital Archives.
64. Race, *UT Southwestern*, p. 11.
65. Today, it is known as UT Southwestern Medical School and is part of the UT Southwestern Medical Center.
66. Newspaper clipping in the collection of the Dr. Ellen Loeb-Katz Collection, 2002.34.80, DHHRM: *The Daily Times Herald*, Dallas, March 2, 1952. He had changed his name from Ladislav Glattstein. See Larry Gladstone papers, United States Holocaust Memorial Museum, https://collections.ushmm.org/search/catalog/irn45907.
67. Commencement Program and Speech, op cit., 1952.
68. Levi-Montalcini, *In Praise of Imperfection*, pp. 121–122 and 135–136.
69. Rohrlich, 'Jewish Lives: Rita Levi-Montalcini', op cit., pp. 27–28.
70. Levi-Montalcini, *In Praise of Imperfection*, p. 139.
71. Ibid., p. 140.
72. Ibid., p. 141.
73. Ibid., p. 142.
74. Donna Jeanne Haraway, *Staying with the Trouble: Making Kin in the Chthulucene. Experimental Futures*. Duke University Press, 2016.
75. Ibid., p. 4.
76. Levi-Montalcini, *In Praise of Imperfection*, p. 143.
77. SOU 1995:10, pp. 224–226.
78. Martin Raberg, 'Fenomenet Halina Neujahr', *Hertha* 2 (1963), p. 23.
79. SOU 1995:10, p. 226.
80. Ibid.
81. Raberg, 'Fenomenet Halina Neujahr', p. 24.
82. 'Working in haematology: 5 Pros and 5 Cons', MyHSN, www.myhsn.co.uk/top-tip/working-in-haematology-5-pros-and-5-cons. Accessed 11 May 2025.
83. Hill, *Blood and Betrayal*, op cit., p. 62.
84. Blair Jones, 'Waits for Cure: Leukemia Story Told in Letter', *The Marshall News Messenger* (Marshall, Texas), 28 December 1958, p. 3.
85. Ricky Ray Hemophilia Relief Fund Act of 1995, op cit., p. 259.
86. F1 AC-4815 Czajkowska, Halina, Rödakorsflykting, Statens utrymningskommission Kanslibyrån (SUK), Riksarkivet (RA), Stockholm, Sweden.
87. Duffin, 'Editor's Epilogue', in Robinson, *Heaven, Hell and Purgatory*, op cit., p. 171.
88. F1 AC-4815 Czajkowska, Halina, SUK.
89. Duffin, 'Editor's Epilogue', in Robinson, *Heaven, Hell and Purgatory*, p. 171.
90. 'In memory of Professor Eva Klein (1925–2025)', Karolinska Institutet, https://news.ki.se/in-memory-of-professor-eva-klein-1925-2025. Accessed 6 June 2025.
91. Hargittai, *Women Scientists*, op cit., p. 48.
92. 'In memory of Professor Eva Klein (1925–2025)', Karolinska Institutet, op cit.
93. Weisse, *Heart to Heart*, op cit., p. 276. Pick, USC interview, tape 5.
94. Gordon, *All Our Lives*, op cit., p. 177.
95. Haraway, *Staying with the Trouble*, op cit., p. 4.

Chapter 6

1. Quoted in Rydberg, 'Forskare mot all odds', op cit., p. 9. Translated from Swedish.
2. Pick, USC interview, tape 4.
3. Marc. A. Silver, Earl N. Silber, David Lieb. 'In Memoriam: Ruth Pick, M.D', *Circulation* 109 (2004): 696–697, p. 697.
4. 'Toward the next century: Heart Disease', *Chicago Tribune* (Illinois), 12 April 1981, p. 7.
5. Pick, USC interview, tape 1.
6. Ruth Bondy, Chaya Naor, trans. *Trapped: Essays on the History of the Czech Jews, 1939–1943*. Jerusalem: Yad Vashem, 2008, p. 44.
7. Pick, USC interview, tape 5.
8. Margaret Carroll, 'A few sappy Chicago love stories. Ruth Pick and Alfred Pick', *Chicago Tribune* (Illinois), 12 February 1986, p. 84.
9. See, e.g. E. S. More and M. J. Greer. 'American Women Physicians in 2000: A History in Progress', *JAMWA* 55, 1 (2000): 6–9.
10. Pick, USC Interview, tape 5.
11. Marc. A. Silver, Earl N. Silber, David Lieb. 'In Memoriam: Ruth Pick, M.D', *Circulation* 109 (2004): 696–697, 696.
12. Stan Donaldson, 'Dr. Ruth Pick, 89, Cardiologist survived concentration camp', *Chicago Tribune* (Illinois), 25 July 2003, p. 11.
13. Silver et al., 'In Memoriam', p. 697.
14. Luigi Aloe, 'Rita Levi-Montalcini and the discovery of NGF, the first nerve cell growth factor', *Archives Italiennes de Biologie* 149 (2011), p. 178.
15. See, e.g. 'Leading Ladies of Science', The *Philadelphia Inquirer* (Philadelphia, Pennsylvania), 26 November 1961, pp. 22–24; Stepp, John W. 'Women Scientists: America's Great Potential', *Evening Star* (Washington, DC), 4 February 1962, pp. 4–6.
16. 'Leading Ladies of Science', *The Philadelphia Inquirer*, p. 22.
17. The article appeared in newspapers all over the country. See, for example: Patricia McCormack, 'Boys Afraid of Bright Ones? Girls Hold Back in Class', New York (UPI), *Philadelphia Daily News*, 20 April 1965, p. 32.
18. See, e.g. Des Jardins, *The Madame Curie Complex*, op cit.
19. Levi-Montalcini, *In Praise of Imperfection*, p. 29.
20. Ibid., p. 35.
21. Ibid., p. 38.
22. See Des Jardins, *The Madame Curie Complex*, op cit.
23. Liversidge, 'Interview: Rita Levi-Montalcini', op cit., p. 72.
24. Ibid. Italics in original.
25. Levi-Montalcini, *In Praise of Imperfection*, pp. 192–195.
26. Liversidge, 'Interview: Rita Levi-Montalcini', p. 72.
27. R. Levi-Montalcini, 'NGF: An Uncharted Route', In: F. G. Worden, J. P. Swazey, G. Adelman (eds), *The Neurosciences: Paths of Discovery*. Boston: I. Birkhäuser, 1992, p. 251.
28. Hargittai & Hargittai, *Candid Science II*, op cit., p. 370.
29. David Brown, 'A Brighter Hope for Tomorrow', *St. Louis Globe-Democrat* (Missouri), 23 March 1963, p. 6.
30. Leif Persson, 'Halina är en teknisk doktor', *Husmodern* 1 (1961), p. 49.
31. Sondra R. Herman, 'Sex-Roles and Sexual Attitudes in Sweden: The New Phase', *The Massachusetts Review* 13, 1/2 (1972), p. 46.

32. SOU 1995:10, op cit., p. 3.
33. Ibid., p. 228.
34. Ibid., p. 220.
35. Neujahr, USC interview, tape 8.
36. Martin Raberg, 'Fran koncentrationslägret till universitetet och tecnologie doctorsgraden', *Vi Kvinnor* 5 (1961), p. 9. The same interview also appeared in other publications.
37. Jerzy Adamczyk and Carlösten Nordmark. 'Hon började sin forskarbana i gettot', *Aftonbladet* (Stockholm, Sweden), 25 April 1971, p. 17.
38. Judith Butler, 'Rethinking Vulnerability and Resistance', in Leticia Sabsay, Zeynep Gambetti, and Judith Butler (eds.), *Vulnerability in Resistance.* Durham: Duke University Press Books, 2016, pp. 12–27.
39. Avishai Margalit, *The Ethics of Memory*, Harvard University Press, 2002.
40. Madeleine Salomon, 'Hon överlevde fem koncentrationsläger', *Expressen* (Stockholm, Sweden), 16 March 1992, p. 21.
41. Katrin Säfström, 'Halina överlevde tyskarnas dödsläger', *Dagens Nyheter* (Stockholm, Sweden), 29 April 1987, p. 6.
42. Duffin, 'Editor's Epilogue', in Robinson, *Heaven, Hell and Purgatory*, p. 172.
43. Robinson, *Heaven, Hell and Purgatory*, p. 170.
44. As noted in Chapter One, it is not clear whether she left the Collip Laboratory in 1955 or 1956, nor whether she retired in 1976 or 1977. Her own accounts differ and no sources have yet been found to determine the exact years. Again, my thanks to Jackie Duffin, Shelley McKellar, and Anne Quirk for their diligence in trying to determine the exact years.
45. Duffin, 'Editor's Epilogue', in Robinson, *Heaven, Hell and Purgatory*, pp. 173–174.
46. Ibid., p. 174.
47. Duffin, *Part II*, op cit., p. 110.
48. Halina Robinson, 'Vinca Revisited – Another Happenstance in the Discovery of Vinblastine', *Biochemistry and Cell Biology* 69 (1991): 581–582.
49. Robinson, 'Vinca Revisited', p. 581. For Noble's article, see Robert L. Noble, 'The Discovery of the Vinca Alkaloids-Chemotherapeutic Agents against Cancer', *Biochemistry and Cell Biology* 68 (1990): 1344–1351.
50. Robinson, 'Vinca Revisited', p. 581.
51. Des Jardins, *Madame Curie*, p. 160.
52. R. L. Noble, C. T. Beer and J. H. Cutts, 'Role of Chance Observations in Chemotherapy: Vinca Rosea', *Annals of the New York Academy of Sciences* 76 (1958): 882–894.
53. Robinson, 'Vinca Revisited', p. 582.
54. Jacalyn Duffin, 'Poisoning the Spindle: Serendipity and Discovery of the Anti-Tumor Properties of the Vinca Alkaloids', *Canadian Bulletin of Medical History* 17, 1 (2000): 155–192. The three-part version in P*harmacy in History* has been cited in previous chapters.
55. 'Robert Noble, MD, PhD', Canadian Medical Hall of Fame, www.cdnmedhall.ca/laureates/robertnoble. Accessed 30 May 2025.
56. 'AACR 2011: Georg and Eva Klein, M.D', American Association for Cancer Research, 4 April 2011, www.youtube.com/watch?v=iH0sM067fjo. Accessed June 6, 2025. My transcription differs slightly from the official transcription, which was made through closed captions. For example, where I heard 'and when it's a lady I'm talking to', the closed caption transcription reads, 'and when there's really I'm talking to'.

57. Hargittai, *Women Scientists*, p. 46; Eva Klein, USC interview, tape 5.
58. Hargittai & Hargittai, *Candid Science II*, p. 428.
59. Eva Klein, USC interview, tape 5.
60. Ibid.
61. Hargittai, *Women Scientists*, p. 47.
62. Ibid., p. 48.
63. Janos Gergely, 'The Ruggero Cepellini Lecturer, 1997', *Tissue Antigens* 49 (1997): 544–545, p, 545.
64. Georg and Eva Klein, 'Doing the right thing', *Free Inquiry* 15, 3 (1995).
65. George and Eva Klein, 'Essay review: The scientist as a poet', *Interdisciplinary Science Review* 25, 4 (2000): 256–260, 256.
66. M. G. Masucci & I. Ernberg. 'Eva Klein (1925–2025)', *Nature Immunology* 26, 527 (2025).
67. Shakno, DHHRM Oral History Interview, 2003, part 1.
68. Elisabeth and Wolford Thompson, 'Among our friends', *The McKinney Examiner* (Texas), 12 March 1964, p. 1.
69. W. W. Bauer, 'Women Physicians No Longer Rare', *Galveston Tribune* (Texas), 16 May 1958, p. 4.
70. Vivien Castleberry, 'Whatever Happened to The Women's Movement?' *D Magazine*, 1 November 1987, www.dmagazine.com/publications/d-magazine/1987/november/whatever-happened-to-the-womens-movement.
71. Ibid.
72. 'B.&P.W. Club', *McKinney Daily Courier-Gazette*, 13 November 1953, p. 8', Rosters, Boxes 2-3, 1930–1997, 'Business and Professional Women's Club of Dallas Records', MA86-2, Dallas History & Archives, Dallas Public Library.
73. Loeb, 'The Memoirs of Ellen Loeb', p. 14.
74. Shakno, introduction to 'The Memoirs of Ellen Loeb', p. 1.
75. 'Waits for Cure: Leukemia Story Told in Letter', *The Marshall News Messenger* (Texas), 28 December 1958, p. 3.
76. Dallas (UPI), 'MD Escapes Gas Chambers: In Gratitude She Has Devoted Her Life to Medical Research Dallas (UPI)', *Republican and Herald* (Pottsville, Pennsylvania), 10 April 1961, p. 9. Elsewhere, the article is attributed to Preston McGraw. In November 1963, McGraw reported on President John F. Kennedy's arrival in Dallas and his subsequent assassination.
77. Obituary, Mort Katz, *Fort Worth Star-Telegram*, 3 September 2003, p. 23; Mort Katz, *Marriage Survival Kit: A Daily Guide to Happier Marriage*, revised and expanded edition, Farnsworth Publishing Co., 1976.
78. *Texas Jewish Post*, 23 July 1964. Also, *Texas Jewish Post*, 4 June 1964.
79. Kay Holmquist, '"Real Selves" Found in Weekly Searches at Group Sessions', *Fort Worth Star-Telegram* (Texas), 27 August 1972, p. 89.
80. 'Group Granted Charter to Raise Funds for Research Against Killer Leukemia', *Texas Jewish Post*, 8 October 1970, p. 4-A.

Epilogue

1. Obituaries. 'Dr. Ellen Loeb Katz … medical chief at cancer unit', *Fort Worth Star-Telegram*, 18 April 1980, p. 40.

2. Carolyn Ondrejas, 'What to do when red and green make blue', *Fort Worth Star-Telegram* (Texas), 14 December 1981, p. 13A.
3. Ellen Loeb, J. M. Hill, N. O. Hill, A. MacLellan, A. Khan, T. R. Alexander, and M. Adachi, 'Treatment of Acute Leukemia with L-Asparaginase', *Experimental and Clinical Effects of L-Asparaginase* (Heidelberg: Springer-Verlag Berlin, 1970): 204–218, 204.
4. Diane Plotkin, Appendix to 'The Memoirs of Ellen Loeb', op cit., p. 46.
5. Silver et al., 'In Memoriam', op cit., p. 697.
6. Cited in Donaldson, 'Dr. Ruth Pick, 89', op cit.
7. Pompian, 'Your Heart', op cit., p. 148.
8. Pick, USC interview, tape 5.
9. Victoria Martínez, 'The Stockholm Holocaust memorial – A restoration of human dignity and a warning against inhumanity', *The Local Sweden*, 2 August 2017. www.thelocal.se/20170803/the-stockholm-holocaust-memorial-a-restoration-of-human-dignity-and-a-warning-against-inhumanity. Accessed 5 September 2025.
10. Halina Neujahr, 'Warszawa – Majdanek – Skarzysko – Norrköping – Bergen-Belsen – Norrköping: Minnen och reflektioner', in Roman Wasserman Wroblewski (ed.) *6 tusen av 6 miljoner ett requiem.* Stockholm: Reproprint AB, 1995, 25–45, here pp. 43–44. Translated from Swedish.
11. Levi-Montalcini, *In Praise of Imperfection*, p. 203.
12. Alison Abbott, 'One Hundred Years of Rita', op cit., p. 565.
13. M. Bentivoglio, 'In Laudatio: Looking at the Future with Rita', *Neuroscience* 252 (2013): 438–442.
14. Rohrlich, 'Jewish Lives: Rita Levi-Montalcini', op cit., p. 51.
15. Moses Chao, Antonino Cattaneo, William Mobley, 'In Laudatio: Rita Levi-Montalcini: The Story of an Uncommon Intellect and Spirit', *Neuroscience* 252 (2015): 421–437.
16. Levi-Montalcini, *In Praise of Imperfection*, p. 5.
17. Michelle Both, 'This Holocaust survivor left a trove of artwork when she died. It's now on display in London, Ont', CBC, 15 August 2023. www.cbc.ca/news/canada/london/this-holocaust-survivor-left-a-trove-of-artwork-when-she-died-it-s-now-on-display-in-london-ont-1.6933664. Accessed 5 September 2025.
18. Duffin, 'Editor's Epilogue', in Robinson, *Heaven, Hell and Purgatory*, p. 175.
19. See 'St. Peter's Cemetery Polish Military Monument', Government of Canada. www.veterans.gc.ca/en/remembrance/memorials/canada/st-peters-cemetery-polish-military-monument#toc-visitor-information. Accessed 6 September 2025.
20. Obituary: James Russell Robinson, O'Neil Funeral Home. www.oneilfuneralhome.ca/obituaries/james-robinson. Accessed 6 September 2025.
21. Robinson, *Heaven, Hell and Purgatory*, p. xiii.
22. Duffin, 'Editor's Epilogue', in Robinson, *Heaven, Hell and Purgatory*, p. 174.
23. Michelle Both, 'This Holocaust survivor left a trove of artwork when she died', CBC, 15 August 2023. www.cbc.ca/news/canada/london/this-holocaust-survivor-left-a-trove-of-artwork-when-she-died-it-s-now-on-display-in-london-ont-1.6933664#. Accessed 6 June 2025.
24. Keri Ferguson, 'Legacy gift of $2M largest ever for Western Libraries', Western News, 3 June 2024, https://news.westernu.ca/2024/06/2m-gift-western-libraries. Accessed 6 June 2025.
25. Anna Bodin, 'Vi möttes på en bräda', *Dagens Nyheter* (Stockholm, Sweden), 28 July 2015, p. 19.

26. 'Cancerforskaren Eva Klein, 90: Arbetet håller mig ung', P4 Stockholm, Sveriges Radio, 11 November 2015, https://sverigesradio.se/artikel/6299920. Accessed 6 June 2025.
27. Masucci & Ernberg, 'Eva Klein (1925–2025)', op cit.
28. 'In memory of Professor Eva Klein (1925–2025)', Karolinska Institutet, https://news.ki.se/in-memory-of-professor-eva-klein-1925-2025. Accessed 7 June 2025.
29. 'Klein Éva', Külhoni magyar tudósportrék, Hungarian Academy of Sciences, 15 November 2014. https://real.mtak.hu/185981/. Accessed 7 June 2025.
30. Klein, USC interview, tape 6.
31. 'Szirmai Jan', Yad Vashem, https://collections.yadvashem.org/en/righteous/5600279. 'Clara Ambrus', www.yadvashem.org/righteous/stories/ambrus.html. Accessed 6 June 2025.

Index